RHS

ROYAL HORTICULTURAL SOCIETY

HOW TO GARDEN

 DK

CONTENTS

6 Introduction

8 ASSESSING YOUR GARDEN
10 How to assess your garden
12 The weather
14 Sun, shade, and the seasons
16 Your microclimate
18 Your soil
22 Garden inspiration

26 FLOWERBEDS & BORDERS
28 Designing beds and borders
30 Size, shape, and layout
32 Choosing your plants
40 Making a planting plan
42 Sunny sites
46 Shady sites
50 Wet, sticky soils
54 Dry, free-draining soils
58 Exposed sites
62 Rock gardens
66 Choosing and buying plants

68 PERENNIALS, ANNUALS & BIENNIALS
70 Growing perennials
80 Growing annuals and biennials

86 BULBS
88 Growing bulbs

98 SHRUBS
100 Growing shrubs
106 Pruning shrubs

118 CLIMBERS
120 Growing climbers
126 Pruning climbers

134 ROSES
136 Growing roses

148 GRASSES, BAMBOOS & FERNS
150 Growing grasses
156 Growing bamboos
162 Growing ferns

166 LAWNS, MEADOWS & GRAVEL GARDENS
168 What sort of lawn?
170 Looking after your lawn
176 What's wrong with my lawn?
180 Making a new lawn
183 Alternatives to grass
184 Wildflower meadows
186 Gravel gardens

192 TREES
194 Garden trees
200 Caring for trees

ROYAL HORTICULTURAL SOCIETY

HOW TO GARDEN

210 HEDGES
212 Garden hedges
218 Caring for hedges

222 CONTAINERS
224 Growing plants in containers
228 Care and maintenance
232 Growing crops in containers

240 EDIBLES
242 Where to grow crops
244 Growing vegetables
248 Root vegetables
254 Leafy vegetables
256 Peas and beans
258 Fruiting vegetables
261 Growing herbs
264 Growing fruit
268 Fruit trees
272 Fruit bushes and soft fruits

276 TOOLS & TECHNIQUES
278 The basic tool kit
282 Watering
286 Mulches
288 Feeding your plants
290 Compost
292 Digging

295 Weeding
296 Eco-gardening
298 Propagating plants

312 GARDEN DOCTOR
314 What's wrong?
316 Garden pests and parasites
320 Plant diseases
324 Keeping plants healthy

328 GARDEN YEAR PLANNER
330 Early spring
331 Mid- to late spring
332 Early summer
333 Mid- to late summer
334 Autumn
335 Winter

336 Glossary
340 Understanding plant names
342 Index
351 Acknowledgments

INTRODUCTION

If you are new to gardening, it may come as a surprise to learn that experienced hands envy you. You are about to embark on an enthralling and satisfying pursuit which may prove as rewarding to you as it is to those who have been gardening for many years.

Gardeners once learnt their skills at their parents' knee, or, perhaps, from friends or neighbours. Today, a mass of books, magazines, television programmes, and websites offer help and advice, but trying to piece together all this disparate information can be time-consuming and confusing.

RHS How to Garden brings together everything you need to know to begin gardening, and breaks the information down into straightforward, easy-to-follow steps. Every page is full of the results of practical experience and there is plenty of sound advice to help with every problem and uncertainty. In this book you'll find a guide through the basics, so you can get to understand your soil and site, with advice on choosing good plants and looking after them properly. Later in the book is a guide to propagation – how to make new plants for free. As your skills, experience, and knowledge increase, this book will remain a valued companion, always available to give a quick and to-the-point answer. In fact, you don't have to be a beginner to find it useful.

However small your garden, you can use a single, carefully chosen container plant to provide a focal point, or in this case a woodland landscape in miniature.

Restricting the colour palette can result in strikingly appealing effects. Here, roses and clematis of similar colours have been chosen for impact.

Making new plants can become an addictive hobby. You can gain a real sense of satisfaction from seeing your own cuttings thrive, and watching apparently lifeless seeds develop into healthy plants.

Watching the garden change with the seasons helps you understand how it has evolved, and gives you the opportunity to assess the pros and cons of all its features.

WHERE TO START?

Possibly you already have a vision of how you want the garden to look. Practical steps you can take to achieve it are outlined, starting perhaps with containers or improving the lawn. Learning at your own pace, work up to flowerbeds and borders, trees, shrubs, and hedges. You won't find recipes for makeovers or grand designs; the aim is to help you make improvements now, and develop the know-how to create the garden of your dreams. Better to start small than risk lots of money in what may turn out to be a discouraging, costly disappointment.

Instead, take the time to look at your garden through the days and months; observation is at the heart of successful gardening, and this book will help you assess what you've got. Established trees, shrubs or hedges can cast deep shade but might provide privacy, shelter, or beautiful features in other seasons with attractive bark, berries, or blossom. You will learn how to renovate them if they need it, and to enhance them with flowering climbers, a scattering of bulbs, or a surrounding border.

GETTING A PLANT'S-EYE VIEW

Choosing plants that thrive in your garden, given its soil, site, and locality, is the key to gardening success. It follows that knowing a little bit about the plants themselves – how they work, what they need to survive, and how they have adapted to different amounts of sun, shade, heat, or drought – will help you match them to your own garden conditions. To this end, the book also includes a brief guide to how plants, soil, and climate work together – enough to help you fathom what is going on behind the scenes. It is equally important to learn to recognize plants and planting combinations that are never going to work in the way you want.

Just an inkling of these amazing, invisible processes can give you that bit of extra confidence so helpful in making the leap of faith involved in choosing plants, sowing seeds, growing your own produce, or pruning a shrub. For example, knowing how plants respond to having stems removed will help you anticipate the consequences of when and where you make a pruning cut. Without this, you are really making a stab in the dark.

Of course there will be setbacks. That (within reason) is part of the fascination. You may discover problems that are of your own making – and how to remedy them – but you can also take comfort from knowing that some things, like the weather, really are beyond your control.

Time spent gardening should be a pleasure. It is often best done in spurts; a relaxed, daily wander through the garden with secateurs and a trowel allows you to deal with minor troubles before they get out of hand, and warns you when more robust remedies are needed. Get in the habit of keeping a notebook handy however; you never know when inspiration will strike!

ASSESSING
YOUR GARDEN

——

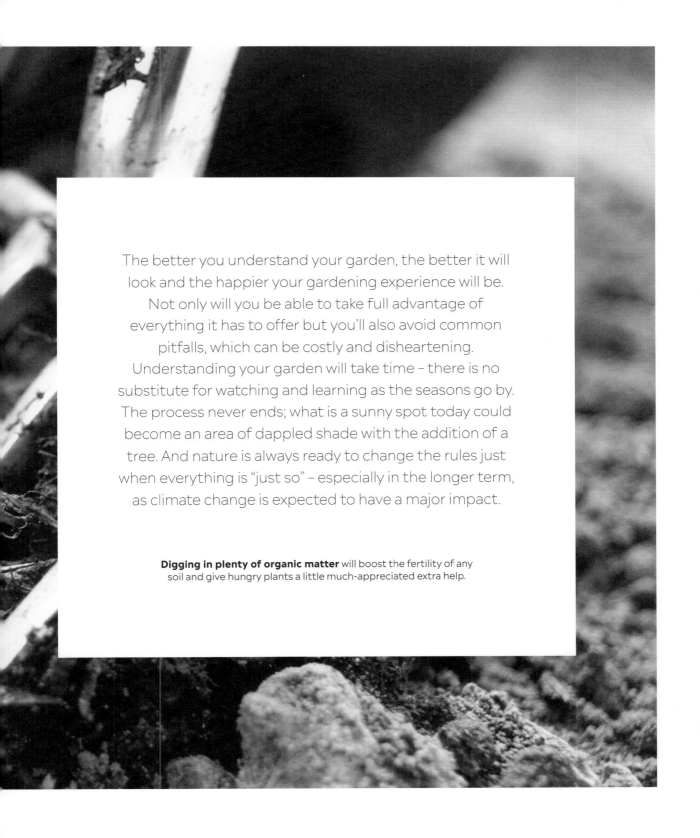

The better you understand your garden, the better it will look and the happier your gardening experience will be. Not only will you be able to take full advantage of everything it has to offer but you'll also avoid common pitfalls, which can be costly and disheartening. Understanding your garden will take time – there is no substitute for watching and learning as the seasons go by. The process never ends; what is a sunny spot today could become an area of dappled shade with the addition of a tree. And nature is always ready to change the rules just when everything is "just so" – especially in the longer term, as climate change is expected to have a major impact.

Digging in plenty of organic matter will boost the fertility of any soil and give hungry plants a little much-appreciated extra help.

HOW TO ASSESS YOUR GARDEN

Brand new gardens can be a bleak prospect, especially if the "topsoil" is little more than a token gesture to cover up builders' rubble, plastic bags, and broken tools. An "inherited" garden can be a double-edged sword, depending on whether it is to your taste or not, and how enthusiastically the previous owners gardened. An old, neglected garden can be just as daunting as a new one – what should stay, and what should go?

THE GARDEN'S PURPOSE

There are a few crucial questions you must ask:

• **Who will use it?** You may love spiky plants and ponds, but if you have a young family then they will probably not be the best choice.

• **What will the garden be used for?** Gardens can be designed for entertaining, as a fun space for children, or simply to show off plants. You might want to encourage wildlife, such as birds or butterflies, into your garden; consider how to balance your needs with that of the wildlife you want to encourage.

• **How much time can you spare?** Work out how much free time you have at the moment, and how much of it you are willing to give up. You don't want your garden to become a millstone rather than a pleasure. The free time you have will inform your plant choice, border sizes, even whether you have a lawn or not.

• **What are your timescale and budget?** It may be better to tackle a small area each year rather than embark on a grand and costly plan. Start with the area you see the most, perhaps laying a patio or redoing a border. Any hard landscaping should be carried out in sections as you can afford them. If you might be moving on, buying fast-growing plants might be a better idea than investing in long-term projects.

Being able to entertain in the garden is important for many people. A patio can easily be adapted into an outdoor eating area by adding a barbeque, a table, and some chairs.

WHAT SORT OF GARDEN HAVE YOU GOT?

What you do with your garden will be affected by whether it is brand new, established, or neglected.

New garden

• **If your soil** is appalling, buy in good quality topsoil from a local source. Reject soil that is dusty, stony, weedy, wet, or "blue" (which is what happens when the air has been squeezed out). Before the soil is unloaded from the truck, check that the pH is close to your own garden's pH.

• **Establish your priorities:** once the plan is complete, decide which elements to establish first. It may be more practical to install a patio, garden shed, and a washing line before a pond and fountain.

Wild garden

• **Think before you act,** as beneath some more wild or natural gardens, treasures are waiting to be revealed.

• **Weeding** on a small scale can be done by hand; if the problem is overwhelming, using chemical weedkillers may be the only option. For an organic approach, clear the area with a strimmer, then cover it with light-excluding tarpaulin or old carpet (this can take a few months to work).

• **Take a clean sweep.** If progressive renovation is out of the question, perhaps the best option is to think of the garden as "new" and consider the steps above.

Mature garden

• **Assess the style.** Reconstructing an established garden can be hard, but putting your own personality on a place is part of the experience of developing and enjoying a garden.

• **Choose what to keep.** Some features may be there for a good reason. A mature tree might conceal a distant tower block or overlooking windows, so consider the existing features with care.

• **Assess the utilities and hard landscaping,** because gardens must be practical as well as beautiful. Take time to ensure you are not removing something that you may need. Conversely, is the garden missing something that you consider essential, like an outdoor tap or space for a barbecue?

(from top) **A new garden, a wild garden, a mature garden**

THE WEATHER

The climatic conditions that affect plants most are rainfall, extremes of temperature, wind strength and direction, and light levels. Knowing the local climate is the first step in understanding which plants grow best in your garden. While most gardens can be manipulated in one way or another to overcome prevailing conditions, the predominant climate – the macroclimate – will always influence plant selection.

THE EFFECTS OF COLD

The length and severity of winter often dictate what you can grow. You can water in dry spells, or boost drainage where excess moisture is a problem, but it is harder to keep temperatures above damaging levels. Hardiness is vitally important, and it is worth understanding the different ways cold can affect plants.

An air frost – when the air temperature drops to 0°C (32°F) while the ground remains unfrozen – is most common in early autumn. A ground frost occurs when the soil temperature falls below freezing; it is more frequent in winter, and can occur while the air is above freezing.

Long winter freezes largely define which plants are hardy in your climate. Less severe frosts tend to happen under cloudless skies on still nights, and are caused by the ground losing warmth by radiating it out into the night sky. If conditions persist, the ground freezes solid. Harder frosts are caused by a moving mass of cold air, with icy winds driving frost deep into the soil. Evergreen plants keep losing moisture from their leaves in winter, especially in windy weather.

In a valley, rivers draw weather systems towards them, and hills can affect rainfall so one side of a valley is often wetter than the other. Cold or freezing air will often settle and persist in valleys.

Coastal gardens suffer few frosts because of the proximity of the sea and the regular sea breezes, although the air is salt-laden and can damage plants.

With the soil frozen, they cannot take up more water; the foliage and green stems are scorched and killed.

SOIL AND MOISTURE

Too much soil moisture in winter or too little in summer is a common theme of gardening life. Walls, fences, and hedges can affect soil moisture by casting "rain shadows". Dig down to discover the depth of the dryness, as this will determine what you plant.

Many plants prefer good drainage and will thrive in dry soil; improve drainage by incorporating plenty of moisture-retentive organic matter before planting. If your garden is very damp, create well-drained spots by building raised beds, or by planting on ridges and mounds. Otherwise, try to choose plants that will tolerate wet conditions.

WHAT IS HARDINESS?

Hardiness is a plant's ability to withstand the local climate. In an area subject to frost, this means resistance to cold; elsewhere it means resistance to heat or drought. Knowing where a plant comes from does not automatically tell you if it will survive in your garden; lavender and rosemary, for example, grow naturally on well-drained land in a warm climate, but will thrive in moister conditions and cooler climates.

Generally, plants may be described as tender, half-hardy, or hardy. For more detailed descriptions, please refer to the table below for the RHS's nine hardiness categories. These serve as a general indicator of a plant's preferred growing conditions, but bear in mind that they are only a guideline, and many other factors may affect overall hardiness.

HARDINESS ZONES

RATING	TEMPERATURE RANGES	CATEGORY	DEFINITION
H1a	>15°C (59°F)	Heated greenhouse – tropical	Under glass all year.
H1b	10–15°C (50–59°F)	Heated greenhouse – subtropical	Can be grown outside in the summer in hotter, sunny, sheltered locations (e.g. city centres), but generally perform better under glass all year round.
H1c	5–10°C (41–50°F)	Heated greenhouse – warm temperate	Can be grown outside in summer throughout most of the UK while daytime temperatures are high enough for growth.
H2	1–5°C (34–41°F)	Cool or frost-free greenhouse	Tolerates low temperatures, but will not survive being frozen. Requires glasshouse conditions except in frost-free areas or coastal extremities. Can be grown outside once frost risk passes.
H3	-5–1°C (23–34°F)	Unheated greenhouse/ mild winter	Hardy in coastal and relatively mild parts of the UK, except in hard winters and when at risk from sudden, early frosts. May be hardy elsewhere with wall shelter or good microclimate. Can often survive with some artificial protection in winter.
H4	-10–-5°C (14–23°F)	Average winter	Hardy throughout most of the UK apart from inland valleys, at altitude and central/northerly locations. May suffer foliage damage and stem dieback in harsh winters in cold gardens. Plants in pots are more vulnerable.
H5	-15–-10°C (5–14°F)	Cold winter	Hardy in most places throughout the UK, even in severe winters. May not withstand open/exposed sites or central/northern locations. Many evergreens will risk suffering from foliage damage, and plants in pots will be at increased risk.
H6	-20–-15°C (-4–5°F)	Very cold winter	Hardy in all of the UK and northern Europe. Many plants grown in containers will be damaged unless given protection.
H7	<-20°C (-4°F)	Very hardy	Hardy in the severest European continental climates, including exposed locations in the UK.

SUN, SHADE, AND THE SEASONS

Because some plants are adapted to a sunny place, and some to shade, finding out where the sun falls in your garden is crucial to success. The sun affects the amount of light and warmth, and therefore which plants you can grow. Take some time to observe the passage of the sun, not just through the course of the day, but throughout the seasons. This will allow you to build a picture of which areas are permanently shady, and which get most sun. Once you are familiar with these areas, you can start to think about which plants will thrive in them, and plan your garden accordingly.

A sunny wall both stores and reflects heat. It can extend the growing season and help wood to ripen in late summer.

ASPECT

First, find out which direction your garden faces – north, south, east, or west. This is known as its "aspect", and will help you to determine whether the back or the front garden receives the most sun. As a general rule, south- or west-facing gardens are warmer and sunnier than north- or east-facing plots.

Start by standing with your back to the house. The direction you are facing, whether it is north, east, south, or west, is the same as your garden. If you are unsure, remember the sun rises in the east and sets in the west. East-facing gardens get the morning sun, while those facing west have sun during the afternoon. South-facing gardens are often bright for much of the day, but north-facing plots don't get much direct sun at all. East- and north-facing gardens can be unsuitable for tender plants, either because there is not enough warmth, or because the temperature rise after the night-time cool is too rapid. Remember though, each wall or fence in the garden also has an aspect, and can create its own sun-trap or shady area.

Where there is dappled shade from open trees, or where the shade from buildings lasts for less than half a day in summer, plants won't grow as much as in the open, but they will be more sheltered, which can make up for the lower light levels.

THE EFFECT OF THE SEASONS

The seasonal variation can be quite marked – the trajectory of the sun varies from summer, when it rises high in the sky, to winter when it hugs the horizon. Clever use of plants can make the most of these changes: the low watery sunlight of midwinter, for example, will illuminate colourful stems, such as those of coppiced dogwoods and willows, and bring interest to the winter garden.

If you live in a town and your garden is surrounded by buildings, the winter sun may barely peep over the rooftops and into the garden, although it will be sunny in summer when the sun moves through the sky directly above.

The look of the garden can change considerably through the seasons. The light quality and brightness changes and patches of shade move. Deciduous trees and shrubs lose their leaves in winter, at the same time as most herbaceous plants die down. It takes care to ensure a garden looks interesting and attractive all year round.

• **In spring,** the bare branches of deciduous trees allow lots of light into the garden; spring bulbs and early flowering perennials such as hellebores and lungworts can take advantage of this. Tree blossom provides a touch of colour.

• **In summer,** shade appears as deciduous trees come into full leaf. In darker areas, grow shade tolerant plants such as heucheras or geraniums. Variegated shrubs, like the mock orange *Philadelphus coronarius* 'Variegatus', or white-flowered shrubs such as hydrangeas, can brighten up gloomy corners.

• **In autumn,** enjoy the leaf colour of deciduous trees before they drop. Light levels are similar to those of springtime as the sun sinks lower, but the quality of the light is often more subdued.

• **In winter,** observe the garden critically – if the structure works now it should work for the whole year. Sculptural evergreens are constant but don't be tempted to include too many or the effect will be spotty. Areas under trees have some protection from the worst effect of frost.

YOUR MICROCLIMATE

Microclimates are local variations in the general climate, and can include soil type, exposure or shelter, and sun or shade. Understanding microclimates enables you to exploit them to your advantage. Even across a small area, there will be enough variation to grow a broader range of plants than you might first think. Microclimates can even be created; for example, when laying a patio, leave a few gaps between paving slabs and fill them with gritty soil to make planting pockets. In a sunny spot, drought-tolerant herbs like thyme or oregano will thrive in the additional reflected heat and light from the paving.

WIND TURBULENCE

Wind is probably the most important factor in the garden environment. Gales and storms can cause substantial physical damage, but even normal breezes have an impact because they dry out leaves and plants will need watering more often. Even in a small garden some areas may be more exposed to wind than others. Wind turbulence is caused by solid walls, fences, and buildings as the wind tumbles over them, creating an eddying effect. You can reduce the problem by using barriers that let some air through. Try to identify the windier or more turbulent parts

Wind damage caused by gusts and eddies swirling round solid barriers is nearly as bad as that of unchecked wind. Semi-permeable barriers, such as trees and shrubs, hedges, trellis, or non-solid walls and fences, provide more effective shelter. They may also cast less shade, and don't completely block your views.

of your garden and reduce the problem by planting hedges or living screens. A barrier to wind can shelter an area of garden behind it equivalent in length to five times the barrier's height.

Wind disturbs the moist air layer on the surface of leaves, making them lose water more quickly than on a still day. As the plants begin to dry out, the leaves close their pores. This prevents carbon dioxide getting into plants, and photosynthesis – and therefore growth – stops. This is why plants grow best in the shelter of hedges and fences.

Shelter belts slow wind speed over some distance, creating a niche for large, leafy plants that can be made ragged or lose too much water in a breeze.

SUN TRAPS

Walls and greenhouses will absorb heat in the sun, then release it slowly later. Walls can also create frost shadows, so placing containers against them during winter will help to keep plants frost-free; a tender azara, for example, will exploit the conditions found in a very sheltered corner of the garden and help brighten up a dark spot. South-facing walls reflect heat, so provide the perfect planting opportunity for less hardy plants. They are traditionally used for growing and ripening fruit like peaches and plums, and for ornamentals such as roses, which may flower earlier due to the extra warmth.

FROST POCKETS

Frost pockets form as chilly air sinks to the lowest point it can. Areas at risk include dips, valley bottoms, and places where cold air can collect behind a barrier. Slopes are generally dry because water runs to the lowest-lying areas in a garden. Lower areas may also be more prone to frost, because cold air sinks. The combination of frost and wet soil is especially damaging, and only fully hardy plants will tolerate such conditions.

(top) **Deep, damp shade** is ideal for lush, leafy plants that dislike high temperatures. Damp shade can either be sheltered and relatively frost-free or, if a barrier such as a hedge or fence is creating a frost pocket, be cold and slow to thaw in winter.

(bottom) **The ground at the foot of a wall** will be sheltered from rain; sun-loving plants, such as these alstroemerias, will thrive there, although they'll need extra care while they establish.

YOUR SOIL

Far from being mere "dirt", soil is the raw material with amazing properties out of which you create your gardening dreams. Plant roots find moisture and food in the spaces between soil particles, and the soil's job is to hold enough moisture to sustain the garden in dry spells, but allow surplus water to drain away in wet weather. It also protects roots from heat and cold. Soil is just a thin skin on the planet's surface, but all life on land depends on it.

Adding mulch to soil will help improve the overall texture and moisture levels, keeping it healthy.

WHAT KIND OF SOIL HAVE I GOT?

Assessing your soil is essential for successful planting; while some plants tolerate a range of soil types, many have definite preferences. Dig an inspection pit; it will help you evaluate your soil in terms of the following factors:

• **Texture** can vary from sandy through to loam, to clay, and all types in between.
• **The depth of the topsoil,** or the uppermost, fertile layer of soil. The deeper the topsoil, the better.
• **The type and depth of subsoil** in your garden might reveal some surprising combinations, such as gravelly soil over clay, or even acid loam over chalk. If your topsoil is shallow, the subsoil is more important, particularly for trees and larger shrubs that need to make deep roots for stability.
• **The moisture content** varies as different soil types have different water-holding capacities; a little rain may drain through sand quickly, while the same amount can make a clay soil quite wet.

Clay soil or sandy soil?

The size of soil particles determines its texture and what grows best in it. Take a handful of moist topsoil, and shape it into a ball. Try to work it into a sausage shape, and rub it between your fingers. The "ideal" soil is loam, which will hold together in a ball and show finger impressions, without being sticky. Most soils, however, tend towards either clay or sand.

Acid soil or alkaline soil?

Your soil's acidity or alkalinity affects what plants you can grow. Some, like rhododendrons, do not grow well in an alkaline soil, while others, such as clematis, thrive on alkaline, or "limy", sites. The term pH is a way of expressing how acid or alkaline a substance is: the lower the pH number, the more acid the soil. Acidity and alkalinity depend on how much calcium is in the soil. Acid soils lack calcium and alkaline ones have it in excess, while neutral soils – 7 on the pH scale – have just enough calcium to mop up acidity.

• **Alkaline soils** have a pH of over 7. You cannot make them acid in the long-term, and acid-loving plants will certainly fail. However, limy soil is often well-drained and quick to warm up in spring. Incorporate plenty of organic matter and a wide range of plants will thrive, but don't overfeed, as this results in tall, floppy plants.

• **Acid soils** have a pH of 1–6. They are ideal for a wide range of plants, but acid-loving or "ericaceous" plants are almost impossible to grow elsewhere. If your soil is not acid, you can grow them in containers filled with ericaceous compost. You can make acid soil more alkaline by adding lime, but think carefully before using it. Most plants will put up with some acidity, and you can grow a wider range in mildly acid soils. In theory, you can add acid material to limy soils to lower the pH, but acidifying materials, such as sulphur dust, act slowly, are needed in very large amounts, and they're expensive.

Clay soil feels smooth and sticky, will roll into a sausage and bend into a loop, and is shiny when rubbed. The tiny particles pack together so water does not easily drain, and air spaces are minute. It is slow to warm up in spring and sticky to work.

Sandy soil feels gritty and falls apart as you try to mould it into a ball. Sand grains are the largest soil particles, with big air spaces through which water can drain freely. Sandy soils warm up fast in spring, but dry out quickly in droughts.

TESTING SOIL FOR PH LEVEL

You can get a good idea of your garden's pH using a testing kit from a garden centre or DIY store. Kits come with full instructions, are quick and simple to use, and are usually more reliable than the meters sold for home use.

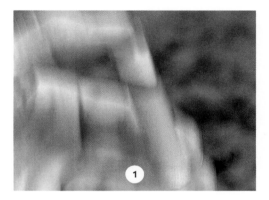

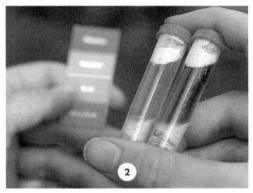

1 Take samples from different parts of the garden, as soil is seldom uniform. Make a note of where each sample comes from so that you can exploit any variations, such as an acid patch, when you draw up your planting scheme. Add the soil to the chemical solution in the test tube and shake to mix.

2 Match the coloured solution against the pH chart to assess what kind of soil you have. Dark green indicates alkaline soil, bright green is neutral, and yellow or orange indicates acid.

TOP PLANTS FOR ACID SOILS

1 *Acer palmatum* 'Sango-kaku'
Coral-bark Japanese maple
A beautiful acer that produces
coral-pink shoots in winter.
The leaves change colour from
yellow in spring, through green in
summer, to orange-yellow in autumn.
H 6m (20ft); S 5m (15ft).

2 *Skimmia japonica* 'Bronze Knight' **Skimmia**
Robust, evergreen shrub with attractive
foliage and dark red buds in winter.
H 1.2m (4ft); S 1.2m (4ft).

3 *Rhododendron* 'Palestrina' **Azalea**
Compact, evergreen azalea with masses
of pure white flowers in late spring.
H 1.2m (4ft); S 1.2m (4ft).

4 *Fothergilla gardenii* **Witch alder**
Small, dense, acid-soil-loving shrub
with white spring flowers and russet autumn
tones, good for a woodland garden.
H 1m (3ft); S 1m (3ft).

5 *Erica vagans* 'Birch Glow' **Cornish heath**
A low-growing, spreading, heathland shrub
with bright, rose-pink flowers.
H 30cm (12in); S 50cm (20in).

6 *Cornus kousa* 'Chinensis' **Dogwood**
A small garden tree that starts to
flower when very young, with a mass
of lime-green, petal-like bracts.
H 7m (22ft); S 5m (15ft).

7 *Gaultheria tasmanica* **Pernettya**
Spreading, mat-like plant with
white flowers in spring followed
by berry-like fruits.
H 8cm (3in); S to 25cm (10in).

8 *Camellia japonica* 'Guilio Nuccio' **Camellia**
Large, elegant, evergreen shrub with
semi-double, rose-red flowers
from late winter to midsummer.
H 9m (28ft); S 8m (25ft).

TOP PLANTS FOR ALKALINE SOILS

1 *Clematis* 'Huldine' **Clematis**
Happy on most soils but especially when the pH is high. Cup-shaped, white and mauve flowers from late summer into autumn.
H 3–5m (10–15ft); S 2m (6ft).

2 *Crocus tommasinianus* 'Ruby Giant' **Crocus**
A reliable, easy-to-grow variety that flowers early and may produce either pale lilac or reddish-purple flowers.
H 8–10cm (3–4in); S 2.5cm (1in).

3 *Photinia davidiana* **Photinia**
Tall evergreen usually grown for autumn leaf colour and bright red, berry-like fruits.
H 8m (25ft); S 6m (20ft).

4 *Caryopteris* x *clandonensis* 'Worcester Gold' **Caryopteris**
Mound-shaped shrub that thrives on chalky soils. 'Worcester Gold' has small, lavender-blue flowers and yellow-green foliage.
H 1m (3ft); S 1.5m (5ft).

5 *Malus* x *purpurea* 'Lemoinei' **Crab apple**
A mid-sized garden tree with dark red-purple leaves and wine-red blossom in spring.
H 8m (25ft); S 8m (25ft).

6 *Hebe* 'Great Orme' **Hebe**
Evergreen shrub with distinctive spikes of deep pink flowers that gradually lose their colour and fade to white.
H 1.2m (4ft); S 1.2m (4ft).

7 *Campanula glomerata* 'Superba' **Campanula**
Vigorous, spreading perennial with round heads of bell-shaped, purple flowers.
H 60cm (24in); S 1m (3ft) or more.

8 *Forsythia suspensa* **Forsythia**
A deciduous shrub valued for its early spring display of bright yellow flowers.
H 3m (10ft); S 3m (10ft).

GARDEN INSPIRATION

Once you've assessed your garden, you can use all of that knowledge to achieve the space that you want. But before you head for the garden centre, it's a good idea to come up with a realistic plan. Try to visualize your dream garden and think about what you want from the space. Do you need a play area for children? Somewhere to entertain family and friends? Here are some ideas to inspire you.

(left) **Mixed flower borders** are often a riot of unstructured colour, combining annuals, biennials and perennials such as these irises, verbascum, achillea, purple *Allium cristophii*, and the poppy Papaver orientale 'Patty's Plum'.

(right) **A raised wooden deck** leads straight out from the open French windows so that the dining area feels as if it is half indoors and half outdoors. Within just a year or two, the grape vine will grow right over the pergola to provide some welcome shade from the sun in the height of summer.

(above) **Two curving granite sett paths** lead to a circular paved area where a wooden bench provides a simple seat. The layout draws the eye naturally through the garden and creates a slightly illusory sense of increased perspective.

(right) **A vegetable plot**, however small, is a valuable addition to any family garden. Few things taste better – or are as good for you – as home-grown vegetables freshly picked from the garden and eaten soon after. At the very least, you should plant a few herbs or patio vegetables within easy reach of the kitchen door.

(above) **Clean, geometric lines** are the hallmark of this modernist design. Carefully placed contrasting areas of wooden decking, stone paving, and raised squares of grass divide the space horizontally, while limestone walls, clipped hedges, and glass screens divide it vertically. Bamboos tower above everything.

(right) **A quiet corner** in a crowded cottage garden is the perfect place to escape and relax. Scented plants such as a fragrant honeysuckle complete the restorative experience.

(above) **A mown grass path** is an inviting and practical route through a summer meadow left to self-seed with a carpet of native wild flowers – including yellow buttercups and white ox-eye daisies.

(above) **Organized chaos** may be the overall effect here, yet there's some artful planning in the choice and siting of plants and antique gardening "props". Yellow giant scabious *Cephalaria gigantea*, the purple-red flowers of *Knautia macedonica*, sisyrinchium, and lupins all give height to the border, while alchemilla, salvia, and thyme spill out over the gravel at the front.

(right) **A hand-woven wicker cage** surrounding a fat ball allows birds access to their food but prevents squirrels from stealing it. You're more likely to be successful in attracting birds if you employ a quiet, slightly overgrown, wild corner of the garden.

(above) **A private, sheltered corner** of the garden can act as an invaluable hideaway from the demands of hectic family life. Surrounding the space with shrubby plants to protect against chilly winds and to provide shade from the midday sun will only increase the welcome sense of seclusion and retreat.

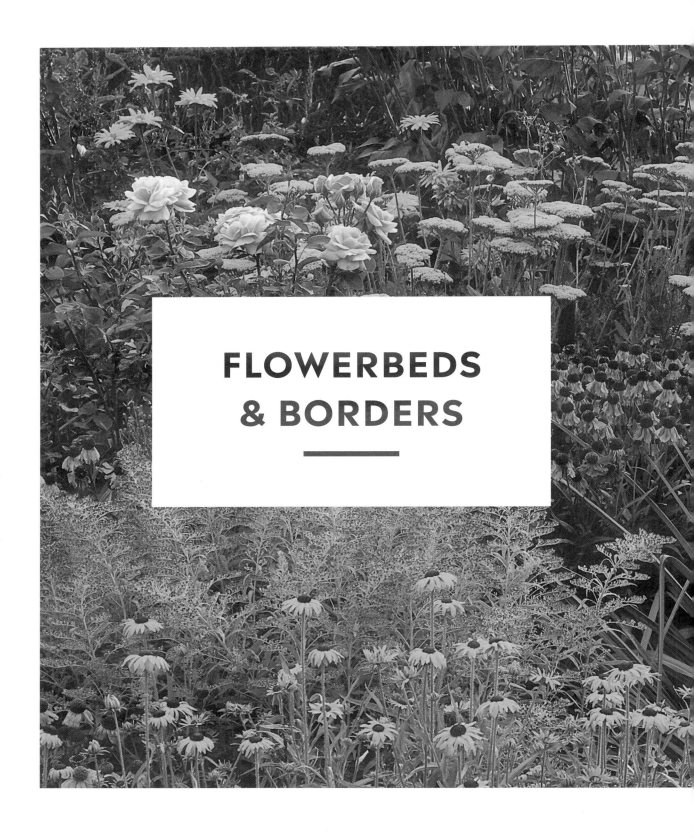

FLOWERBEDS
& BORDERS

Strictly speaking, a border is a planting bed that runs along a boundary line, such as a fence, hedge or wall, although traditionally, borders were often used to enclose formal parterres or lawns. Beds, on the other hand, are freestanding, and frequently used as focal points in the centre of a lawn or gravel garden. Formal gardens, such as parterres, use geometric-shaped beds, often edged with low, evergreen hedging, and set out in a symmetrical pattern. For many centuries, beds have been used for growing vegetables, herbs and flowers, and still offer a practical solution for vegetable plots, with plants on all sides receiving sun throughout the day.

Mixed borders depend for their success on careful planning. The effect may appear effortless, but layered planting such as this requires knowledge of when different plants come into flower and how large they will grow.

DESIGNING BEDS AND BORDERS

The first step when planning your garden – whether you are faced with an overgrown jungle, or a pretty plot that needs only minor adjustments – is to note down what you want from your space. Does your dream garden include beds and borders packed with colourful plants, an activity area for children, somewhere to entertain family and friends, or a combination of all three? You also need to consider practicalities: where will the dustbins live, do you need a bike shed, how much space is required for the dining table?

Over winter, leave herbaceous plants to stand, rather than cutting back the dead growth. Their dried seedheads and stems will take on a magical quality when dusted with frost or snow.

GARDEN STYLE AND STRUCTURE

The style of garden you choose will affect the size, shape and position of beds and borders, as well as the hard landscaping. Even if you don't want an ordered garden, introduce a basic structure so that your garden looks interesting in the winter. Consider planting a tree, and outline beds or borders with bricks, paving slabs, clay tiles, or a low hedge, to help define and structure the space. The harsh outlines will soon soften once the growing season starts.

Take time to plan your planting and you will be rewarded with interest all year round. Evergreen trees and shrubs provide structure in winter; mix these with perennials that retain attractive seedheads in winter, such as sedums, achilleas, and echinops. Although their winter display is not colourful, they create interesting shapes and prevent borders from looking flat and lifeless. Perennials also provide a wonderful spectacle on frosty mornings, and draw birds into the garden to feast on the seeds and insects hibernating inside empty pods.

Grasses offer invaluable interest in autumn and winter gardens. The dried flower spikes of some, such as *Stipa gigantea* and molinias, stand well into autumn, while miscanthus and calamagrostis remain upright until spring.

A formal garden will have lines of uniform planting. These low evergreens have been pruned into smart balls, and the foliage of berberis and sage, rather than flowers, injects colour.

In an informal garden, plants may billow freely over the path. Grasses and herbaceous perennials provide a colourful display all summer, and background shrubs and trees offer year-round structure.

PLANTING STYLES

Your choice of plants should reflect the overall style you want to achieve:

• **Traditional herbaceous borders** consist only of herbaceous perennials, and will perform from early spring into autumn. In the early 20th century, herbaceous borders were often only on show for six to eight weeks in summer, while themed areas, such as iris gardens, dahlia borders, and autumn gardens, catered for the other seasons. Few of our small, modern gardens can accommodate such themed garden rooms, and we need our borders to perform for as long as possible.

• **Mixed beds and borders** consist of several different types of plants. Although perennials play a key role, they are assisted by shrubs and roses, and may also include annual flowers. If the border is backed by a wall or trellis, the height of the display can be increased with a backdrop of climbers. Alternatively, create height in a border by adding a tripod for climbers.

A mixed border offers both summer flowers and structure in the winter, but it has its drawbacks. Woody plants have spreading roots that take water and nutrients away from perennials, and shrubs may shade out smaller plants; compensate by planting compact shrubs that have smaller root systems.

• **Annual plantings** can be labour-intensive, but provide a great summer show. As many flower over a long period, the overall effect can be much more impressive than a herbaceous border, where plants come and go as the season progresses. When autumn arrives, though, the bed or border will be empty for a few months. The ground will have to be dug and prepared each spring, before any seeds can be sown.

STYLES OF BED

Beds can be informal or formal, depending on their shape and design. Those that are part of a formal layout or parterre are usually geometric in shape and neatly edged, while informal beds have curved outlines and less defined edges.

• **Formal beds** are typified by neatness and symmetry. Planting is kept in its allotted space with walls of closely cropped hedging, timber strips, steel bands, clay tiles or even glass bottle bottoms.

• **Informal beds** have softer outlines and can be positioned to lead the eye to a focal point.

SIZE, SHAPE, AND LAYOUT

The planting of a border is like choreography: you need to choose and position the star plants as well as their supporting cast. You also need to take account of which plants flower when, and decide which colours should be brought together and which kept apart (*see pp.36–37*). Although this can take time, it is also good fun. Even if you don't get it completely right the first time, you can take comfort in the fact that you could change it next year. When you are caught up in the excitement of planting – usually in spring or autumn – it is easy to forget to include plants that may look uninspiring at the time but will provide a good show in another season. Therefore plan ahead carefully, and ensure that your beds and borders perform well throughout the seasons.

Large swathes of colourful planting in broad, curved borders have a dynamic effect on designs. Find out when your chosen plants flower, and plan a succession of colour all year.

SPACINGS AND FLOWERING TIMES

Small flowering plants, such as aubrietas and doronicums, perform the spring show in the garden. As the year progresses, larger flowering plants dominate. Many of the real giants, such as rudbeckias, do not come into their own until late summer. Although it is tempting to site all the small plants near the front and the tall ones at the back of a border, the result can be rather dull. It is also harder to get an even distribution of flowers through the season. To avoid this, bring a few tall, airy plants, such as grasses or the slender-stemmed *Verbena bonariensis*, closer to the front. They will make the overall shape of the display more interesting, while allowing you to see the plants behind.

When planning, allow enough space for each perennial plant to grow to its full height and spread. This may leave gaps between plants in the first year, but it is better than planting too closely and creating an overcrowded bed or border. Try covering the bare soil with hardy annuals in the first year; this should prevent it drying out and becoming overcome with weeds.

The style in which you plant borders and beds will be influenced by the type of plants you prefer. If you opt for mixed planting, consider planting shrubs or small trees for year-round structure. Use stakes or a broom to mark their intended positions, to get an impression of how they affect the rest of the garden.

GIVING SHAPE TO BEDS

If you grow perennials in an open bed rather than against a wall or hedge, they won't need to compete for moisture or light, and rarely need staking. Informal island beds have little architectural presence, so create structure through planting. Use bold, architectural feature plants, such as New Zealand flax or *Euphorbia characias*.

Although most borders are formal rectangles that cling to the boundaries, the front edges can be curved to create an informal look. If you have a pair of facing borders, you could weave a path down the middle, and allow the borders to fill the spaces on either side. To allow easy access in summer when planting flows over the edges, ensure the path is at least 1.5m (5ft) wide. An S-shaped pathway also creates planting spaces of varying depths. In smaller gardens, this provides a few wide areas for trees, large shrubs and swathes of perennials.

Whatever your garden style, borders under 1m (3ft) wide can be very limiting; any less than this, and you will be continually battling against plants breaking their boundaries and shading out lawns or covering pathways.

Leafy plants, such as ornamental grasses and hostas, offer structure with their long season of interest, and help to disguise the unsightly dying foliage of earlier-flowering specimens.

CHOOSING YOUR PLANTS

Flowering plants are traditionally classified according to their life cycle. Some live and die during the course of a single year, flowering and producing seeds that, with luck, will germinate the following year. Some have a two-year life cycle. And others are able – in theory, at least – to live indefinitely, year after year. This is where the terms annuals, biennials, and perennials come from.

Cotoneaster (shrub)

PLANT TYPES AND CHARACTERISTICS

• **Annuals** germinate, produce leaves, stems, and flowers, set seed, and die – all within a year. If the plant can withstand frost (a hardy annual), its seed may sprout almost immediately and grow into a young plant before winter. Other seeds lie dormant, ready to germinate in the following spring.

• **Biennials** germinate and grow into good-sized plants in the first year (some, such as foxgloves, or *Digitalis*, produce rosettes of leaves). It is not until the second year that a biennial goes on to produce flowers and set seed. After this, the entire plant dies down and its life cycle is complete.

• **Perennials** regrow each year from an underground rootstock. They produce leaves, flowers, and seeds before dying down at the end of the season. The following spring, new shoots appear and the cycle begins again.

• **Bulbs** are also perennials, but their fleshy roots and sometimes their growth habits are different from those of garden perennials. With bulbs are included corms, rhizomes, and tubers (*see pp.86–97*).

• **Shrubs** are woody-stemmed plants. They may be evergreen or deciduous and are often grown for their foliage rather than flowers (*see p.100*). Most shrubs are bush-like and grow from a number of different stems at their base.

(opposite, clockwise from top left) **Pot marigolds** (annual), **foxgloves** (biennial), **daffodils** (bulb), **sea holly** (perennial)

Easy-care plants, such as the rudbeckias and daylilies shown here, have sturdy stems that don't need staking. Daylilies' strappy leaves also provide lush cover when early summer-flowering plants are dying back.

BEDDING PLANTS OR ANNUALS?

Traditionally, bedding plants are used in geometric, formal display beds, and are stripped out and replaced at the start of each season. Therefore for winter and spring you could grow winter-flowering pansies, followed by bellis daisies, forget-me-nots, wallflowers, and spring bulbs, such as grape hyacinths and tulips. For the summer, try any of the many tender bedding plants, such as busy Lizzies, heliotropes, lobelias, marigolds, pelargoniums, and *Salvia splendens*.

Although this system provides a great colour show for many months of the year, as well as allowing you to experiment with different planting schemes, it is labour-intensive. Twice a year, the bed has to be cleared, dug, and replanted. You also need to buy a large number of plants each time at considerable expense, or raise some of them in advance in a frost-free greenhouse.

Easier and cheaper are hardy annuals. You can grow these from seed, which you can scatter straight on to the spot where you want them to flower (*see pp.80–83*). The main drawback is that you may have to wait 10–16 weeks for the flowers to bloom. Success rates can also be unpredictable – numerous factors influence germination, such as age of the seed, soil moisture and temperature. The seedlings are also vulnerable to weather damage, and some pests and diseases.

Spring bulbs such as daffodils and alliums (*below*) bloom for many years before you will need to lift and divide them. Alliums have sturdy stems that don't need staking and handsome, long-lasting seedheads.

LOW-MAINTENANCE PLANTS

If you choose your plants wisely, you can make life easier and reduce the time you need to spend maintaining them. Stick to sturdy herbaceous perennials and grasses, and avoid using annuals or bedding plants. Supplement the perennials with some small evergreens and ground-cover shrubs, such as euonymus, heathers, or tiny-leaved creeping cotoneasters, in mixed beds and borders to provide year-round greenery.

Don't grow plants such as delphiniums and double peonies that need staking. Instead, look for easy-care alternatives. For example, monk's hoods, like delphiniums, are great for punctuating a border with their tall flower spikes in shades of blue and purple. But unlike delphiniums, they are not eaten by slugs, and stay upright without staking.

Every garden has sites with different growing conditions, or microclimates (*see pp.16–17*), so finding plants that perform well may involve some trial and error. Don't be afraid to move a plant if it fails in one spot; it may well thrive elsewhere. Once you have planted a bed or border, keep it free of weeds by mulching it regularly (*see pp.286–287*).

You can even design your garden with low-maintenance plants and still enjoy colour and interest throughout the year (*see below*).

Summer is the highlight of the flower-garden year. For a low-maintenance display, fill the borders with showy plants that reach for the skies. *Nepeta sibirica* has masses of flowers on self-supporting, tall stems.

Autumn flowers include Michaelmas daisies and chrysanthemums. Also keep an eye out for colourful berries or seedheads, such as these Chinese lanterns (*Physalis alkekengi*) with their fiery orange seed capsules.

Winter blooms are precious commodities since few flowering plants brave the cold. Violas (*below*) are among the most reliable, and perform best in a warm spot, such as on a patio. Hellebores also bloom in winter.

CREATING COLOUR SCHEMES

Flowers and plants offer an exceptional colour palette that would make any artist envious. The subtlety in shades and hues, further enhanced by the different textures of the leaves and petals, creates a vast range of colours to play with. Remember that plants are not the only source of colour in a garden, and it is also important to consider fences, walls, paving, and containers, all of which add to the final effect.

For some, the huge choice of colour in the garden offers an exciting design opportunity, but for others, choosing colour schemes can be nothing short of daunting. Many different colour theories are used by garden designers, but if you are nervous that your plot may end up a frenzy of clashing hues, or a poor palette of dull shades, the colour wheel offers a helpful guide (*see opposite*).

Opposing colours

The colour wheel is made up of primary colours (red, yellow, and blue), and secondary colours (orange, green, and purple); the latter are formed when the two primary colours on either side are mixed together.

These primary and secondary colours provide the foundation for successful colour schemes. You will see that each primary colour sits opposite a secondary colour. These "opposing colours" are complementary and work extremely well when used together in a garden context. Thus, red goes well with green, yellow with purple, and blue with orange.

Bearing in mind that most foliage is green, and that – on a good day, at least – the sky is blue, it is difficult to be strict about this theory, because the majority of colours in the garden go well with blue and green. However, it is undeniable that blue and orange do combine very well, and that yellow and purple create a pleasing match.

Adjoining colours

Opposing colours create visual excitement, but can be overbearing. Adjoining colours, which sit side-by-side on the colour wheel, create more subtle combinations. These were favoured by Gertrude Jekyll, who created many amazing gardens in the early 20th century. She divided the colours into two categories: cool colours, consisting of blues, purples, lilacs and pinks, and hot colours, which

Using opposing colours guarantees an eye-catching display. This simple but effective combination of fiery orange cosmos and purple-blue morning glories works especially well.

include yellows, oranges and reds. Jekyll saw that it was important to use the in-between shades to create subtle blends. Rather than planting only yellows and reds, fuse the two with shades of orange and some plants with yellow and red in their foliage to carry the colour theme throughout the year.

Monochrome planting

If you are disciplined enough, restrict your palette to just one colour. Try creating a sunny yellow border, or a cool silvery-white one. It may seem straightforward to plant a monochrome garden using plants of the same colour, but it is important to try to create variation and contrast within a scheme, as without that it will appear bland and monotonous.

For example, when using white, do not restrict yourself to pure white flowers, but also include cream, lime green, pinkish and mauve-tinted whites. Likewise, choose the foliage of the plants carefully, and include pale to dark greens, grey, silver, and blue-tinted leaves. When opting for yellows or reds, go for the full tonal spectrum within those colours, and think about how to mirror your colour choice in the accompanying foliage.

Green

This is the most dominant and important colour in the garden and there are myriad shades of green to choose from. For colour that will outlast a flowering display, mix foliage in various shades of lime, apple, and blue-green.

THE COLOUR WHEEL

A colour wheel is an at-a-glance guide to which colours work well together. It is very easy to use and takes the guesswork out of creating successful colour combinations within the garden.

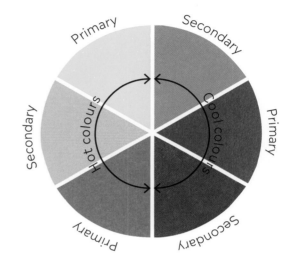

Monochrome schemes can be monotonous unless you include interesting foliage colours. Here, cream and white flowers are mirrored by silver and yellow variegated hostas and grasses.

CHOOSING PLANTS FOR YEAR-ROUND COLOUR

Certain plants will inject colour into the garden all year round. They provide structure and a sense of permanence as everything else changes day by day. Some are suitable for containers as focal points, while others will enhance a colourful display in a border.

White
- *Artemisia stelleriana* 'Boughton Silver'
- *Euonymus fortunei* 'Silver Queen'
- *Hebe pinguifolia* 'Pagei'
- *Betula utilis* subsp. *jacquemontii* 'Silver Shadow'

Purple
- *Ophiopogon planiscapus* 'Kokuryu'
- *Salvia officinalis* 'Purpurascens'
- *Ajuga reptans* 'Atropurpurea'
- *Phormium tenax* Purpureum Group

Blue
- *Festuca glauca* 'Elijah Blue'
- *Leymus arenarius*
- *Juniperus squamata* 'Blue Star'
- *Ruta graveolens* 'Jackman's Blue'

Green
- *Fatsia japonica*
- *Hedera hibernica*
- *Pittosporum tobira*
- *Stipa gigantea*

Yellow
- *Choisya ternata* Sundance
- *Ilex aquifolium* 'Golden Milkboy'
- *Carex oshimensis* 'Evergold'
- *Hebe ochracea* 'James Stirling'

COLOUR THROUGH THE SEASONS

One of the most effective ways to use colour is to allow the seasons to dictate the palette. When spring arrives, nature seems so relieved that the dreary weather is over that it responds with a burst of sunny colours and every imaginable shade of green. As the season progresses into early summer, the colour palette calms down a little, concentrating on clean, fresh colours, such as the blues of delphiniums and geraniums, and white and pink roses and peonies.

As summer heats up, so do the colours: warm yellows, burnt oranges, and velvety reds signal hot, sunny days. The fading light that heralds autumn is reflected in the subdued tones of asters. Winter's colours, while more subtle, are just as effective – white snowdrops never fail to cut through the gloom.

(opposite, clockwise from top left)
Spring sees an abundance of sunshine yellows and vibrant golds typified by narcissi.

Late summer flower colours hot up before autumn with amber heleniums and golden crocosmias.

Autumn colour is provided by bright berries and fruits, and the rich reds, oranges, yellows, and browns of dying leaves.

Winter needn't be dreary with ice-white snowdrops and cheery pink cyclamen to look forward to.

MAKING A PLANTING PLAN

Be generous when marking out the area for your bed or border.
If you can afford the space, allow at least 3m (10ft) to provide a good
planting width. You can then build up several layers of plants, one behind the
other, and use shrubs and climbers to create height towards the back. If the choice
is between two narrow borders and one deep one, you'll find the larger more
rewarding, however intimidating it may seem at first.

Key plants, such as tall wispy grasses, add focal points to borders and beds. Structural
foliage plants are useful inclusions as their impact is more enduring than transient flowers.

CHOOSING THE KEY PLANTS

Working from an accurate base plan makes designing a new bed or border easier, and helps you decide how many plants you need to fill it. When selecting plants, think of your border as a stage performance, and pick out the leading cast. This should include plants that have an important presence because of their handsome shape, decorative leaves, or a long flowering season. Good candidates include colourful foliage shrubs, such as berberis and cotinus, and many of the grasses, sedums and euphorbias. Repeating these every few metres brings rhythm into a scheme.

Think of where you would like to see your stars: sedums flower in autumn and are good front-of-the-border plants, while taller grasses, such as molinia or calamagrostis, are better placed in the middle or background. Mark the plants on your plan with a circle representing their final spread. Most of the larger perennials will spread to 45cm (18in), while smaller ones will reach 30cm (12in); check plant labels for the spreads of shrubs, and note that their size can be affected by your soil and garden conditions.

PLANNING IN COLOUR

Once the key plants are on your plan, choose back-up players that offer more subtle effects. Keep a note of their flowering times and colours, and, if in doubt, make an overlay with tracing paper, on to which you can mark the colour of each plant using different pen types or symbols. Do a few overlays for the different seasons to show when your chosen cast is in flower, where the gaps are, and at what time of year they need to be filled.

CONTRASTING FLOWER SHAPES

Think of the contrasts in flower shape as well as colour, height and season. Plants with spike-like flowers, such as veronicas, delphiniums, foxgloves, and verbascums, or upright, linear foliage, such as irises or daylilies, will introduce a vertical accent. Those with round disc-like flowerheads, such as sedums, yarrows, fennel, and all the cow-parsley relatives, create a horizontal plane. Mixing these two types of flower paints an exciting picture and can help you to create repetition and symmetry throughout the garden if this is the look you want. Distinctive spherical shapes, such as alliums and *Echinops ritro*, add to the structure and interest.

(right, from top) **Combine different flower shapes** – such as foxgloves, yarrow, and echinops – in beds and borders to create an architectural feast for the eye.

SUNNY SITES

Sun-lovers tend to be among the most free-flowering of all garden plants. If you are lucky enough to have a sunny garden with reliably moist soil – from a natural water course, perhaps – you will be able to grow a huge range of plants. But more often, sun combines with little rain and you find that the soil dries out in summer. On a dry site, the trick is to select those plants that can cope with a lack of moisture and don't need constant watering, which is time-consuming and costly both to the environment and to your pocket.

A dry garden has been created at Hyde Hall, a Royal Horticultural Society garden. It is a large hillside garden that evokes a rocky, Mediterranean slope by using silvery, spiky-leaved plants that weave in and out of large boulders. Despite its size, it contains plenty of inspiration for domestic gardeners.

DROUGHT-TOLERANT PLANTS

In addition to sun, plants that can survive without much water require the following:

• **Good drainage:** few drought-tolerant plants will put up with a waterlogged soil in winter. As a general rule, it is cold combined with wet that kills plants, not cold alone.

• **Low nutrient levels:** overfeeding encourages fast growth of soft, sappy foliage, which wilts in hot weather. Plants grown in poor soils also flower more profusely and for longer.

• **Good air circulation:** this helps reduce fungal diseases and rotting at the plant's "neck" (the point at which the plant emerges from the soil).

Often, a garden may have the right climatic conditions for these plants but unsuitably wet, heavy soil. In this case it is essential to improve the drainage so that plants are able to survive the winter months. Incorporate plenty of sandy grit in the topsoil – down to a depth of 25cm (10in) or so – or make raised beds using stone, brick, or timber. Build the retaining walls at least 40cm (16in) high, add a 10cm (4in) layer of 20mm gravel in the bottom and fill to planting level with gritty topsoil. Individual plants can be planted onto raised mounds up to 30cm (12in) high and 1m (3ft) across. This is especially effective for trees and shrubs.

Drought-tolerant, sun-loving plants are ideal for a variety of planting styles including sunny beds and borders, and they will often thrive in containers on a patio. While classic Mediterranean plants, including herbs like lavender and rosemary, are often used in formal knot gardens and parterres, they are probably at their best in informal, naturalistic styles. Use gravel paths to draw the eye through the garden, as they wind among mounds of bright flowers and silvery foliage that are interspersed with aromatic herbs and shrubs.

RECOGNIZING SUN-LOVING PLANTS

The best plants to use in a sunny garden are those that would naturally be found in exposed positions with high light levels, low rainfall, and poor, fast-draining soils. Plants that can cope in such locations have evolved by developing certain characteristics. A sunny garden planted with suitable plants will be much easier to establish and maintain than one with poorly chosen plants.

(clockwise from top left)

• **Silver or grey leaves** reflect heat and sunlight. Many silvery plants, particularly artemisias such as the *Artemisia alba* 'Canescens' shown here, are superb foliage plants.

• **A low-growing** habit helps reduce the drying and damaging effects of wind in exposed places. This rose root (*Rhodiola rosea*) also conserves water with its fleshy leaves, silvery green foliage, and a thick, water-storing root.

• **Small leaves,** such as those of this rock rose, *Helianthemum nummularium*, reduce the surface area through which moisture can be lost. Some shrubs, such as Spanish broom (*Genista hispanica*), go so far as to barely have leaves at all.

• **Waxy leaves,** like those on the pretty, white-flowered *Cistus salviifolius*, are coated with a thick waterproof layer to slow down moisture loss and reduce leaf scorch in very hot weather.

TOP PLANTS FOR SUNNY SITES

1 *Berberis thunbergii* f. *atropurpurea* 'Helmond Pillar'
Barberry
Will grow happily in partial shade but autumn
leaf and berry colour is stronger in full sun.
H 1.5m (5ft); S 60cm (24in).

2 *Eryngium* x *tripartitum* **Sea holly**
Thistle-like flowers, midsummer to early autumn.
Use in a sunny gravel garden.
H 60–90cm (24–36in); S 50cm (20in).

3 *Crocosmia masoniorum* **Montbretia**
Long, arching stems produce vivid, bright
red-orange flowers in midsummer.
H 1.2m (4ft); S 8cm (3in).

4 *Stipa arundinacea* **Pheasant's tail grass**
An evergreen whose new-season dark green
leaves become streaked with orange from
summer through to winter.
H 1m (3ft); S 1.2m (4ft).

5 *Santolina chamaecyparissus* **Lavender cotton**
Scented, grey-green foliage and bright yellow
flowerheads. Grow on dry soils in full sun.
H 50cm (20in); S 1m (3ft).

6 *Pelargonium* 'Apple Blossom Rosebud'
Pelargonium
Rounded clusters of bicoloured pink and white
flowers. Will grow in partial shade, but prefers full sun.
H 30–40cm (12–16in); S 20–25cm (8–10in).

7 *Phlomis cashmeriana* **Phlomis**
Evergreen shrub with sage-like leaves and hooded,
lilac-purple flowers.
H to 90cm (36in); S 60cm (24in).

8 *Papaver rhoeas* Mother of Pearl Group
Field poppy
Sun-loving annuals that self-seed freely. A wide
range of colours.
H to 90cm (36in); S to 30cm (12in).

9 *Aster frikartii* 'Wunder von Stäfa' **Aster**
A sun-lover with violet-blue, daisy-like flowers in late summer and early autumn.
H 70cm (28in); S 45cm (18in).

10 *Hibiscus syriacus* 'Lady Stanley' **Hibiscus**
Deciduous shrub with exotic, pink and white flowers; plentiful in hot, sunny summers.
H 3m (10ft); S 2m (6ft).

11 *Anthemis punctata* subsp. *cupaniana* **Anthemis**
Low-growing, spreading perennial with white, daisy-like flowers, ideal for rock gardens.
H 30cm (12in); S 90cm (36in).

12 *Tulipa* 'Ancilla' **Tulip**
Small tulip with pink blooms flushed red at the petal base. Flowers in mid-spring. A good rock-garden plant.
H 15cm (6in).

13 *Salvia patens* **Salvia**
A perennial with hooded flowers from midsummer to mid-autumn.
H 45–60cm (18–24in); S 45cm (18in).

14 *Yucca filamentosa* 'Bright Edge' **Yucca**
Clump of sharp leaves with yellow edges and tall spires of white flowers. Needs full sun.
H 75cm (30in); S 1.5m (5ft).

15 *Cytisus* x *praecox* 'Allgold' **Broom**
Pale yellow-flowering shrub that will tolerate poor soils but requires full sun. Good for dry borders and rock gardens.
H 1.2m (4ft); S 1.5m (5ft).

16 *Cordyline australis* 'Albertii' **New Zealand cabbage palm**
A palm-like, evergreen shrub that can grow to the size of a small tree. In frost-prone areas, grow in a container and provide shelter in winter.
H 3–10m (10–30ft); S 1–4m (3–12ft).

SHADY SITES

You may see a shady area as having less potential than a sunny one, but there are plenty of plants that thrive in shade. Shade is seldom at a constant level throughout the day and across the seasons, so observation is essential. In a north-facing garden, the shade might be deep and constant close to the house, but less so further away from it. Perhaps your garden is actually quite sunny, but a group of trees have created areas of shade. As always, plants that are adapted to the conditions will be the most successful, and are easily spotted (*see opposite*). Reliably moist soil in shade will open up opportunities for species that would naturally grow at the edges of streams and ponds, while very dry shade will require a different, more limited, range of plants.

Moist shade is enjoyed by many plants. They will do even better if you improve the soil, and perhaps thin out some tree branches so the shade is less dense.

IMPROVING THE SOIL

The main problem in a shady garden tends to be the soil conditions. Trees growing in and around your garden require tremendous amounts of water and nutrients, and can leave the surrounding soil dry and barren. That said, if you are lucky enough to garden on the edge of old woodland, your soil will be enriched by leaf litter. Shade cast by a building, depending on the aspect, can result in soil that is either dry and thin, or dank and structureless.

Soil pH (see p.19) can be affected by trees. A large pine tree, for example, can sometimes acidify the soil around it as its needles fall and rot down; this will affect the plants you can grow there. Nevertheless, all of these soil problems can be overcome with a little know-how and effort.

As most shade-tolerant plants originate from woodland environments, it makes sense to try to recreate similar conditions in your own garden by incorporating as much organic matter – composted bark, home-made leafmould and garden compost (see p.292) – into the soil as possible. This can take time, especially under trees where roots prevent digging in compost to any depth. Here it is best to lay the compost on the surface as a mulch, and keep applying it every spring and autumn so it breaks down slowly into the soil, or is drawn down by earthworms.

LETTING IN MORE LIGHT

Reducing very dense shade vastly increases the number of plants you can grow. You can improve light levels in a number of simple ways.

Shade from buildings can be reduced by whitewashing the walls in order to reflect extra light into the garden.

Shade from dense tree canopies can be dealt with in the following ways:
• **Thinning,** which removes alternate branches.
• **Pollarding,** which involves regularly cutting back the main branches to the trunk or stem.
• **Coppicing,** which means the tree is cut down to ground level to promote multiple, vigorous and decorative new shoots.
• **Crown lifting**, in which the lowest branches are removed to raise the height of the canopy.

RECOGNIZING SHADE-LOVING PLANTS

As with sun-loving plants, shade-tolerant species have adapted over millennia to the conditions in which they grow. You can recognize shade-lovers by looking for the following traits:

Flowering in spring or autumn makes the most of the relatively high light and high rainfall when deciduous trees are leafless. Many bulbs use this strategy; they make the most of the short window of opportunity, flowering and replenishing their food stores, then spending the rest of the year dormant beneath ground.

Dark pigmentation in leaves allows plants to photosynthesize more effectively when they are subjected to low light conditions. Many ericaceous (acid-loving) woodland plants, including camellias, have very dark green, glossy leaves.

Large leaves help to maximize the surface area through which light is absorbed. The leaves of this evergreen bergenia are also thick and leathery, helping to reduce water loss.

Densely packed leaves increase the light-collecting surface area. Some ferns, such as *Polystichum setiferum*, are also covered in hairs to trap moisture, so they can establish and survive in comparatively dry soil.

TOP PLANTS FOR SHADY SITES

1 *Hyacinthoides non-scripta* **Bluebell**
The classic English bluebell is ideal for growing in shady borders.
H 20–40cm (8–16in); S 8cm (3in).

2 *Epimedium* x *perralchicum* **Barrenwort** or **Bishop's mitre**
Evergreen perennial flowering in spring. Will grow in shade beneath trees.
H 40cm (16in); S 60cm (24in).

3 *Hedera helix* 'Glymii' **Ivy**
Most green-leaved ivies grow in shade, if the soil is kept moist. Variegated ivies tend to need more light.
H 2m (6ft).

4 *Pulmonaria saccharata* **Bethlehem sage**
Spreading, ground-cover plant with white-flecked, green leaves and red-violet, violet, or white flowers in spring.
H 30cm (12in); S 60cm (24in).

5 *Hosta* 'Buckshaw Blue' **Hosta**
Slow-growing variety with deep blue-green leaves. Hostas thrive in damp shade – though watch out for slugs and snails.
H 35cm (14in); S 60cm (24in).

6 *Trillium grandiflorum* **Trillium**
Vigorous perennial bearing pure white flowers in spring and summer.
H to 40cm (16in); S 30cm (12in).

7 *Hydrangea paniculata* 'Praecox' **Hydrangea**
Large, deciduous shrub with clusters of open, white flowers from midsummer.
H 3–7m (10–22ft); S 2.5m (8ft).

8 *Convallaria majalis* 'Albostriata' **Lily-of-the-valley**
Beautifully scented shade-lover ideal for growing with ferns in a cool spot.
H 23cm (9in); S 30cm (12in).

9 *Tiarella cordifolia* **Foam flower**
The tiny flowers are borne over a long period from spring to summer. Spreads freely.
H to 3m (10ft); S to 2m (6ft).

10 *Helleborus orientalis* **Lenten rose**
Winter-flowering, in shades of purple, pink, and greeny white. Will tolerate a damp, shady woodland garden.
H 45cm (18in); S 45cm (18in).

11 *Lamium maculatum* 'Album' **Dead nettle**
Low-growing, spreading, nettle-like plant that thrives in shade. Good ground cover among shrubs or in woodland areas.
H 15cm (6in); S 60cm (24in).

12 *Fuchsia* 'Mrs Popple' **Fuchsia**
A bushy, fast-growing fuchsia that should be as happy in light shade as in full sun.
H 1m (3ft); S 1m (3ft).

13 *Viburnum tinus* 'Eve Price' **Viburnum**
A dense, evergreen shrub with clusters of white flowers followed by blue-black fruits.
H 3m (10ft); S 3m (10ft).

14 *Luzula nivea* **Woodrush**
A tufted, grass-like perennial, found in the wild on moorland, bogs and in woodland. Useful for damp shade.
H 2m (6ft); S 2.5m (8ft).

15 *Lamprocapnos spectabilis* **Bleeding heart**
A perennial with dangling, heart-shaped flowers. Grows well in partial shade and sun.
H to 1.2m (4ft); S 45cm (18in).

16 *Geranium phaeum* **Cranesbill**
A good plant for damp shade. Late spring and early summer flowers, deep purple-black, maroon or white.
H 80cm (32in); S 45cm (18in).

WET, STICKY SOILS

Wet, sticky, clay soils can be among the hardest to work with: it is all too easy to destroy their structure and leave behind a gooey, unworkable mess. However, there is no need to be downhearted over this, because clay can be the most rewarding of all soils if prepared well and planted with things that appreciate its moist, fertile nature.

CULTIVATING WET, STICKY SOIL

Working wet soils is best attempted when the soil is at its driest, assuming it dries to some degree at some point in the year. This can mean cultivating at unusual times of the year, such as midsummer, and planting in early rather than late autumn when the soil is still warm and comparatively dry. It may even result in summer cultivation followed by spring planting.

Soil that is always moist can support many specialized plants that would be impossible to grow otherwise. If you have a large, wet area then you could take your cue from nature and create a water meadow garden. This will not only make life easier, but it is also a good way to attract wildlife.

A raised bed is a good option where conditions alternate between wet and dry, making cultivation difficult. It needs to be at least 40cm (16in) high and can be built from timber, brick, or stone. Fill it with topsoil that has been improved with organic matter (or grit for clay soils) to improve the drainage.

There are advantages to this approach, as it makes sure that, rather than sitting in wet soil through the winter, young plants can start actively growing and producing new roots as soon as they are put in the ground. How much cultivation is needed depends on how wet the soil is, and to what depth – this is easily determined by digging inspection pits at varying times of the year and in different locations. Armed with this knowledge you can then decide whether it is worth working with the prevailing conditions, or trying to manipulate them.

DAMP AND DROUGHT

Soil that fluctuates from very wet in the winter to severely dry in the summer is ideally drained with a ditch, soakaway, or buried land drains that carry water. Where this is impractical soils can be improved by incorporating about two bucketfuls of organic matter per square metre to enhance winter drainage. This also helps the soil hold water through dry spells, so your moisture-loving plants don't suffer in summer. Individual plants, especially trees and shrubs, can be planted on mounds of soil or, for hedges, raised ridges. This is an old technique to prevent water naturally settling where the soil has been disturbed – which often results in a newly planted specimen immediately drowning in awater-filled pit.

RELIABLY MOIST SOIL

If you are lucky enough to have a soil that always remains moist, the opportunity to grow huge, lush bog plants is just too good to miss. Incorporating a little grit or light compost may aid structure to some degree, and the application of thick, dry mulch, such as chipped bark, is certainly worthwhile as it will help to retain the moisture but also allow you to walk on the soil when you need to weed or replant – useful when it comes to weeding or replanting. In very damp gardens, consider constructing decks and raised boardwalks to provide seating areas and paths. You can then enjoy your garden all through the year without getting muddy feet or compacted soil.

HOW PLANTS TOLERATE WET SOIL

Moisture-loving and marginal plants are suitable for reliably wet soils because they have adapted in various ways so they don't drown in wet, anaerobic (airless) conditions. Moisture-loving plants thrive in the kind of moist soil found by ponds, streams, and rivers. Marginal plants are similar, but prefer to grow in shallow water, or reliably wet soil that is regularly submerged.

Air channels that run from shoot to root allow oxygen to circulate around bog plants like arum lilies, whose roots would otherwise be starved of oxygen.

Shallow roots are typical because there is no need to search for water – bog buttercups, for example, spread across the soil surface.

Huge leaves are characteristic of wetland plants such as rodgersias. Due to the fact that there is no shortage of water it doesn't matter if they lose it rapidly.

Special equipment is used by the swamp cypress; it can live in water because it develops breathing apparatus on its roots. They push up through the water surface and take in oxygen so the submerged roots can breathe.

TOP PLANTS FOR WET, STICKY SOILS

1 *Ajuga reptans* 'Burgundy Glow' **Bugle**
Justly renowned for tolerating poor, damp soils and full shade.
H 15cm (6in); S 60–90cm (24–36in).

2 *Aconitum* 'Bressingham Spire' **Aconite** or **Monk's hood**
A tall perennial with deep violet flowers that likes cool, moist soil and partial shade.
H 1m (3ft); S 30cm (12in).

3 *Rosa* L'Aimant **Rose**
A reliable, easy-to-grow rose with scented flowers and a tolerance for clay soils.
H 1m (3ft); S 75cm (30in).

4 *Astilbe* x *arendsii* 'Fanal' **Astilbe**
Perennial with feathery, plume-like dark crimson flowers that will grow in fertile clay soils as long as they stay moist in summer.
H 60cm (24in); S 45cm (18in).

5 *Lysimachia ciliata* 'Purpurea' **Loosestrife**
A perennial whose natural habitat is damp grassland or woodland, often near water.
H 1.2m (4ft); S 60cm (24in).

6 *Galanthus nivalis* Sandersii Group **Snowdrop**
Common snowdrops like heavy soils, as long as they don't dry out, and partial shade.
H 10cm (4in); S 30cm–1m (1–3ft).

7 *Eupatorium purpureum* **Joe Pye weed**
Pink, pinkish purple or creamy white flowers are borne from summer to early autumn.
H 1.5m (5ft); S 30cm (12in).

8 *Carex pendula* **Sedge**
Most sedges are found naturally in bog, moorland or damp woodland, so should thrive in moist, clay soils.
H to 1.5m (5ft); S to 1.5m (5ft).

9 *Cotoneaster horizontalis* **Fishbone cotoneaster**
An easy-to-grow shrub that should tolerate a heavy,
clay soil and partial shade.
H 1m (3ft); S 5m (15ft).

10 *Digitalis purpurea* **Foxglove**
A shade-loving biennial native to woodland.
Purple, pink, or white flowers.
H 1–2m (3–6ft); S to 60cm (24in).

11 *Monarda* 'Mahogany' **Bergamot**
Hardy perennial with ragged, wine-red flowers,
tolerant of most soil types but needs to be kept
moist in summer.
H 90cm (36in); S 45cm (18in).

12 *Persicaria affinis* 'Donald Lowndes' **Persicaria**
Low-growing evergreen perennial that produces
bottle-brush spikes of flowers.
H to 20cm (8in); S 30cm (12in).

13 *Camassia leichtlinii* 'Semiplena' **Quamash**
Semi-double, creamy white flowers. Should grow
well in clay soils to which grit has been added for
better drainage.
H 60cm–1.2m (24in–4ft); S 10cm (4in).

14 *Caltha palustris* **Marsh marigold**
Prefers bog conditions or very shallow water.
Waxy yellow flowers in spring.
H 10–40cm (4–16in); S 45cm (18in).

15 *Rodgersia pinnata* 'Superba' **Rodgersia**
A perennial ideal for growing near water, or in damp
woodland. Horse-chestnut-like leaves and tall
clusters of flowers.
H to 1.2m (4ft); S 75cm (30in).

16 *Aruncus aethusifolius* **Aruncus**
Small, compact perennial that positively likes
moist, heavy soils. Creamy white flowers appear
in early summer.
H 25–40cm (10–16in); S 40cm (16in).

DRY, FREE-DRAINING SOILS

By late summer, free-draining, sandy soils are often very short of water, and many plants just won't survive. Some plants, however, will thrive in these trying conditions (*see opposite*). Trees and shrubs, including conifers, are also good choices for these soils. Once established, their deep root systems make them pretty much drought-proof, although they will need watering for the first year or two until they have settled in. Nevertheless, even when you are growing well-adapted plants, it is still worth adding an organic soil improver to help soil structure and stability, without restricting drainage. Composted bark, garden compost, and spent mushroom compost are all good choices.

Many aromatic herbs come from the Mediterranean hillsides and enjoy dry, well-drained soils. They often have attractive foliage and pretty flowers, in addition to their culinary uses.

SANDY SOILS

Light, sandy soils are easy to work, but water runs freely through them, washing out nutrients. Adding plenty of organic matter adds nutrients and helps soak up and retain moisture. Applying a general-purpose fertilizer in spring will also top up nutrient levels. Rose fertilizer is a good choice because it tends to be rich in magnesium, an element often in short supply in sandy soils. Calcium is also easily washed out, so sandy soils are often acid. You can easily check this with a test kit (*see p.19*). In severe cases you may need to add lime every five years.

Sandy soils are also prone to compaction. Again, plenty of organic matter is the answer; dig it in before planting whenever the soil is dry enough to work. Lawns on sandy soil are especially liable to suffer from speedy drainage and compaction. Autumn spiking, feeding, top-dressing with organic matter, and scarifying will be needed to keep them looking good (*see pp.172–173*).

CHALKY SOILS

Chalky or limestone soils are also very free-draining, but are rich in calcium. Don't even think about trying acid-loving plants – camellias, for example – as they will never really thrive, if they survive at all. Other plants, even those adapted to chalky soil, will greatly benefit from mulching, watering, and feeding. Organic matter, added frequently, is vital, although needs to be used generously because it breaks down very quickly in lime-rich soils.

The best turf grasses prefer mildly acid soil so lawns on chalky soil are seldom quite as good as on other types. Consider going for wildflower-rich mixes or eco-lawns, where clover in the lawn keeps it green in dry weather and adds nitrogen, saving the need for feeding.

HOW PLANTS TOLERATE DROUGHT

Plants that naturally grow in dry environments, or on well-drained soils have several ways of coping with the conditions. Nevertheless, a bit of extra care will work wonders. Water plants very thoroughly before and after planting, and keep them just proud of the ground to leave room for a generous mulch (*see pp.286–287*), which will help retain soil moisture.

Thick, fleshy leaves have large cells capable of holding moisture over a long period. *Agave americana* is a classic example, hailing from the dry deserts of the southern United States.

Hairy leaves work by trapping a layer of moist, still air close to the leaf surface. The pasque flower, *Pulsatilla vulgaris*, has softly hairy leaves and is available in many colours, such as the red 'Rubra'.

Long, thick tap roots, for example those of verbascums, yuccas, and this *Eryngium* x *oliverianum*, reach deep into the soil in the quest for water, and store moisture and starch for lean periods.

Summer dormant plants retreat below ground to avoid heat. Many spring bulbs, including tulips, developed this strategy to avoid the heat of the Mediterranean summer. Some other types of plant, like *Geranium tuberosum*, do the same.

TOP PLANTS FOR DRY, FREE-DRAINING SOILS

1 *Allium schoenoprasum* 'Forescate' **Chives**
A pink-flowering chive that likes heat, sun, and fertile, well-drained, sandy soil.
H 60cm (24in); S 7cm (3in).

2 *Cotinus coggygria* Purpureus Group **Smoke bush**
A deciduous shrub that produces masses of tiny flowers and grows well on light, chalky soils.
H 60cm (24in); S 7cm (3in).

3 *Scabiosa lucida* **Scabious**
Small perennials that occur naturally in dry, sunny meadows and on rocky slopes.
H 20cm (8in); S 30cm (12in).

4 *Artemisia* 'Powis Castle' **Mugwort**
Drought-tolerant perennial with silvery grey foliage that prefers light, dry soils.
H 60cm (24in); S 90cm (36in).

5 *Nandina domestica* **Heavenly bamboo**
A tall, flowering shrub that likes light, dry soils, but must not be allowed to dry out. Small white flowers in midsummer.
H 2m (6ft); S 1.5m (5ft).

6 *Penstemon* 'Apple Blossom' **Penstemon**
Perennial grown for foxglove-like flowers. Prefers well-drained soils.
H 45–60cm (18–24in); S 45–60cm (18–24in).

7 *Sempervivum ciliosum* **Houseleek**
A dense, spreading succulent for growing in a rock or gravel garden, in crevices in walls, or in containers.
H 8cm (3in); S 30cm (12in).

8 *Lavandula angustifolia* 'Munstead' **Lavender**
A blue-purple-flowered variety. Lavender thrives in well-drained, sandy soil in full sun.
H 45cm (14in); S 60cm (24in).

9 *Coreopsis grandiflora* 'Badengold' **Coreopsis**
A mass of deep golden-yellow, daisy-like flowers that, if deadheaded, will continue all through the summer.
H to 90cm (36in); S 45cm (18in).

10 *Hebe* 'Bowles' Variety' **Hebe**
A compact flowering shrub that will tolerate poor and slightly alkaline soils if kept moist but well-drained.
H 45cm (18in); S 60cm (24in).

11 *Bergenia* 'Sunningdale' **Elephant's ears**
Glossy-leaved evergreen that produces clusters of bright lilac-magenta flowers from early spring onwards.
H 30–45cm (12–18in); S 45–60cm (18–24in).

12 *Verbascum* 'Cotswold Queen' **Mullein**
Verbascums will grow on most fertile soils but need good drainage and like full sun.
H 1.2m (4ft); S 30cm (12in).

13 *Salvia* 'Blue Spire' **Perovskia**
A plant that tolerates poor, dry, chalky soils. It has silvery foliage and violet-blue flowers.
H 1.2m (4ft); S 1m (3ft).

14 *Weigela* 'Looymansii Aurea' **Weigela**
An easy-going, pink-flowering shrub that will grow in almost any soil but prefers partial shade.
H 1.5m (5ft); S 1.5m (5ft).

15 *Centaurea cyanus* **Cornflower**
With a natural habitat of meadows and grasslands, cornflowers will grow in moist, well-drained soils.
H 20–80cm (8–32in); S 15cm (6in).

16 *Pennisetum alopecuroides* **Fountain grass**
A bristle-flowered, evergreen grass that prefers light, well-drained soils and is ideal for prairie-style, perennial planting.
H 0.6–1.5m (2–5ft); S 0.6–1.2m (2–4ft).

EXPOSED SITES

Like humans, plants are predominantly composed of water. This makes them prone to dehydration – a constant threat in an exposed garden, whether by the coast, or inland, or at a high altitude. Wind can test plants to the limit, and only those species that are adapted to resist its drying effects – either by storing water in their leaves or roots, or reducing the loss of water through their leaves – stand a chance. On the positive side, exposed gardens usually have high light levels and excellent air circulation, reducing fungal disease. But strong sun can also fade subtle flower colours, and good air circulation can quickly turn to an eddying wind that damages plants.

WIND DAMAGE AND STUNTING

Exposure causes slow, stunted growth – a tree in an exposed garden can be as little as half the size of one of the same age and species in a sheltered garden. Trees and shrubs can also be "pruned" by the prevailing wind. In an exposed inland garden, very cold winds are usually the main enemy. If you garden on the coast, you should benefit from a mainly frost-free climate because sea temperatures rarely drop below freezing, and so keep the winds coming over them warm. On the down side, salt spray borne on the breeze can "fry" leaves and often kill plants as effectively as frost.

USING PLANTS FOR SHELTER

Minimize the effects of wind by using a shelter belt of plants. Unlike solid walls and fences, which can cause wind to accelerate and eddy through a garden, plants slow it down gently. A shelter belt can be anything from a hedge to a dense planting of trees and shrubs. One of the best hedging plants for seaside gardens is salt-tolerant *Griselinia littoralis*, while hornbeam (*Carpinus betulus*) makes a good, dense hedge for exposed inland gardens.

You can also use man-made windbreak materials. These usually take the form of a closely woven fabric mesh, which can be fixed onto regularly spaced posts.

Coastal gardens enjoy a mild climate but must endure strong winds and salt spray. Low-growing species can cope especially well, and will not interfere with stunning views.

Roses often thrive in exposed sites because the good air circulation helps prevent fungal diseases and general damp. Choose varieties with strong flower-colours, such as 'Hertfordshire', that will resist fading in high light levels.

Shelter belts or hedges filter and slow wind very effectively to protect other plants in the garden. Choose plants of an appropriate size for your garden – pine trees will grow quite large, whereas roses are more compact.

A double layer will provide greater protection. A windbreak like this will slow the wind considerably over a short distance. The mesh can be removed as soon as the plants are settled – between three and five years. Individual plants can be cosseted in a similar way with a "cage" of mesh netting on posts, and trees or shrubs planted as small "whips" (usually one to two years old) can be given a head start if surrounded by a biodegradable plastic guard to create a warm and sheltered microclimate.

PLANTS FOR EXPOSED GARDENS

Successful seaside gardens often take a cue from nature by using native plants that proliferate on the shoreline. These can provide a backdrop for more exotic ornamentals. Alternatively, provide shelter for plants that don't cope so well with exposure to the elements. Thick wind-reducing plantings on the boundaries of the garden, and a series of "garden rooms" surrounded by more hedges within, create very sheltered areas. In a mild coastal climate, this will enable you to grow tender, exotic plants. The Royal Horticultural Society garden at Hyde Hall (see p.42) is an example of an exposed inland garden, perched on top of a hill among the rolling, arable landscape of mid-Essex. Throughout summer, the prevailing wind is a hot, dry south-westerly that can desiccate plants in no time, while in winter a cold north-easterly can cause significant wind-chill. Despite this, the garden is bursting with plants, the majority of which have been selected because of their tolerance to exposure.

TOP PLANTS FOR EXPOSED SITES

1 *Buddleja davidii* 'Fascinating' **Butterfly bush**
Large, fast-growing, deciduous shrub whose lilac-pink flowers are irresistible to butterflies.
H 3m (10ft); S 5m (15ft).

2 *Prunus spinosa* 'Purpurea' **Blackthorn or Sloe**
A hardy tree or shrub with sharp spines that makes a good hedge. Its edible black fruits are favoured by both birds and gin drinkers.
H 5m (15ft); S 4m (12ft).

3 *Crambe maritima* **Sea kale**
A perennial native to coastal areas with tough, leathery leaves and large clusters of small, white flowers. Stems can be eaten as a vegetable.
H 75cm (30in); S 60cm (24in).

4 *Juniperus squamata* 'Meyeri' **Juniper**
An evergreen, coniferous shrub with blue-green foliage. Originally from the Himalayas, so tolerant of extreme weather.
H 4–10m (12–30ft); S 6–8m (20–25ft).

5 *Hertia cheirifolia* **Othonna**
Evergreen shrub with fleshy leaves and daisy-like yellow flowers in summer.
H 60cm (2ft); S 90cm (3ft).

6 *Rosa* 'Hertfordshire' **Rose**
Ground-cover rose, with large clusters of flat, carmine-pink blooms, freely produced from summer to autumn.
H 45cm (18in); S 1m (3ft).

7 *Helichrysum italicum* **Curry plant**
Small, evergreen shrub with strongly aromatic, silvery grey leaves.
H 60cm (24in); S 1m (3ft).

8 *Griselinia littoralis* **Griselinia**
Good as single plants but also useful as hedging or a coastal windbreak. Vigorous.
H to 8m (25ft); S 5m (15ft).

9 *Tamarix ramosissima* **Tamarisk**
Deciduous shrub with feathery leaves and tiny pink flowers. Withstands winds and sea spray.
H 5m (15ft); S 5m (15ft).

10 *Potentilla fruticosa* 'Red Ace' **Shrubby cinquefoil**
A compact, deciduous, flowering shrub that is fully hardy and tolerant of poor soils.
H 1m (3ft); S 1.5m (5ft).

11 *Argyrocytisus battandieri* **Pineapple broom**
Pineapple-scented flowers in mid- to late summer.
H 5m (15ft); S 5m (15ft).

12 *Festuca glauca* 'Elijah Blue' **Blue fescue**
Tufted, evergreen grass, with spikelets of blue-green flowers in summer.
H to 30cm (12in); S 25cm (10in).

13 *Glaucium flavum* **Yellow horned poppy**
Branched grey stems of golden yellow or orange poppy-like flowers, in summer.
H 30–90cm (12–36in); S to 45cm (18in).

14 *Hippophae rhamnoides* **Sea buckthorn**
Silver-green, deciduous shrub often used for windbreaks, hedges, and to stabilize sand dunes.
H 6m (20ft); S 6m (20ft).

15 *Salvia rosmarinus* **Rosemary**
Best-known as a kitchen herb but can grow into large, resilient shrubs tolerant of poor soils.
H 1.5m (5ft); S 1.5m (5ft).

16 *Agrostemma githago* **Corncockle**
A fast-growing annual that requires full sun and flowers best on poor soils.
H 60–90cm (24–36in); S 30cm (12in).

ROCK GARDENS

Traditionally, rock gardens have served as self-contained areas in which gardeners can grow the sort of plants normally found in high, mountainous, alpine habitats. For this reason, they are purposely constructed from rocks and stones in an attempt to mimic the bare outcrops, scree slopes, ledges, and crevices found at high altitudes. Most of the plants for which such environments are created are small and compact, and tend to be tough and drought-resistant. Many of them flower profusely, in a range of brilliant colours. Contemporary garden design, however, has broadened the definition of the term rock garden. Drawing inspiration from Mediterranean and desert landscapes, and employing lots of hard surfaces such as paving, pebbles, shingle, and gravel, modern rock gardens are as likely to feature cacti and succulents as true alpines.

CREATING A ROCK GARDEN

The best site for an alpine rockery is an open, sunny, south-facing slope. Almost all rock plants grow best with lots of light and in well-drained soil. They don't do well where the soil is damp and heavy, or where there are overhanging trees. If you're constructing a rock garden from scratch, first lay a foundation of coarse rubble, such as broken bricks, to create a gradient or form a mound. Cover it with inverted grass turves or a sheet of polypropylene punctured to let water drain through freely. Place your rocks on top, positioning the largest ones first. Salvaged or second-hand stones are best – particularly sandstone, limestone, or tufa. Aim for a natural-looking mix of bare outcrops, miniature ravines and gullies, and plenty of crevices for planting. Infill the stones with good garden soil, and then spread a top layer of specially prepared compost comprising equal parts loam, garden compost or leaf mould and sharp sand or grit. You may want to leave some areas as scree beds, covering them with a mix of coarse grit, gravel, or stone chippings.

Many rock-garden plants are low-growing and deep-rooted. They form neat clumps or spread horizontally in a mat-like fashion. Most have small leaves to minimize water loss due to evaporation.

The Alpine pink *Dianthus alpinus* 'Joan's Blood' bears magenta-pink flowers in summer.

The rusty-back fern *Asplenium ceterach* is a dwarf variety that rarely grows taller than 15cm (6in).

CHOOSING PLANTS FOR ROCK GARDENS

Some rock-garden plants are true "alpines", and some aren't. Those that are actually come from alpine regions, where they grow above the tree-line, having evolved to withstand the extreme weather conditions found at high altitudes. Most are dwarf or low-growing and hug the ground to avoid buffeting from strong winds. Their leaves tend to be small, fleshy, and often covered by woolly hairs in order to help them retain moisture that would otherwise evaporate in hot sun or constant wind.

Non-alpines include dwarf trees and shrubs, plants that originate on cliffs and shores in coastal areas, a wide range of miniature bulbs, and even cacti and succulents. In general, all share the same dislike of being waterlogged and a preference for free-draining, moderately fertile soils.

Marjoram or oregano is a herb well-suited to rock gardens. *Origanum amanum* has funnel-shaped, pink flowers.

TOP PLANTS FOR ROCK GARDENS

1 *Limnanthes douglasii* **Poached egg plant**
Bright flowers from summer to autumn. Once established, it will self-seed freely.
H 15cm (6in); S 15cm (6in).

2 *Campanula poscharskyana* **Campanula**
Low-growing perennial that produces masses of star-shaped, violet flowers.
H to 15cm (6in); S to 60cm (24in).

3 *Armeria maritima* **Sea thrift**
Seaside plants with grass-like foliage and small rounded flowerheads.
H to 20cm (8in); S to 30cm (12in).

4 *Aubrieta deltoidea* **Aubretia**
Often seen cascading over walls, aubretia is one of the earliest and most colourful spring-flowering perennials.
H 5cm (2in); S 60cm (24in) or more.

5 *Silene schafta* **Campion** or **catchfly**
A useful rock-garden plant that does not flower until late summer and into autumn.
H 25cm (10in); S 30cm (12in).

6 *Thymus* 'Bressingham' **Thyme**
A ground-hugging, aromatic, evergreen shrub that will spread naturally in crevices.
H 15cm (6in); S 45cm (18in).

7 *Gentiana septemfida* **Gentian**
A small, spreading perennial grown primarily for the intense blue or purple-blue flowers that appear in late summer.
H to 15–20cm (6–8in); S to 30cm (12in).

8 *Sedum spathulifolium* 'Cape Blanco' **Stonecrop**
Bright yellow, star-shaped flowers above a mat of rosette-shaped, silver-green leaves.
H 10cm (4in); S 60cm (24in).

9 *Saxifraga sancta* **Saxifrage**
Tufts of bright green leaves from which clusters of small yellow flowers appear in spring.
H 5cm (2in); S 20cm (8in).

10 *Narcissus* 'Tête-à-tête' **Daffodil**
A dwarf variety that produces up to three golden yellow flowers per stem early in spring.
H 15cm (6in).

11 *Gypsophila repens* 'Dorothy Teacher'
Baby's breath
A compact, alpine perennial with blue-green foliage and tiny flowers that are pale pink when new but darken over time.
H 5cm (2in); S to 40cm (16in).

12 *Viola* x *wittrockiana* Crystal Bowl Series **Pansy**
Compact plants in a wide range of clear colours. Summer-flowering.
H to 12cm (5in); S to 30cm (12in).

13 *Anemone blanda* 'Violet Star' **Anemone**
A low-growing perennial with relatively large flowers that are a deep, dark violet colour.
H 15cm (6in); S 15cm (6in).

14 *Geranium cinereum* 'Ballerina' **Geranium**
A dwarf hardy geranium that produces attractive, red-veined pink flowers with dark centres from late spring onwards.
H to 15cm (6in); S to 30cm (12in).

15 *Ajuga reptans* 'Multicolor' **Bugle**
A small, creeping perennial with dark bronze-green leaves splashed with cream and pink.
H 15cm (6in); S 60–90cm (24–36in).

16 *Juniperus communis* 'Compressa' **Juniper**
Slow-growing dwarf conifer whose restrained height keeps it in proportion with other rock-garden plants.
H to 80cm (32in); S 45cm (18in).

CHOOSING AND BUYING PLANTS

If bewildered by the choice of plants available, look for the tried-and-tested star performers from each group of plants. These usually have shapely forms, plenty of flowers, or show good disease resistance, and grow reliably in most gardens. A good guide to the choicest plants is the Award of Garden Merit, or "AGM", bestowed by the Royal Horticultural Society. Some plants are trialled in RHS gardens over several years, and experts decide which plants are especially garden-worthy. If a plant has an AGM, it is usually noted on its label.

WHERE TO BUY YOUR PLANTS

Your local garden centre can be a good place to buy plants, since most have a wide range of well-grown stock, but be aware of possible pitfalls. Firstly, garden centres tend to put plants in full flower on prominent display, to entice gardeners into impulse buying. This is where research comes in handy. Stick to your list as much as possible and resist these temptations – you may find you have no space or no suitable site or soil for a hastily purchased plant. Many garden-centre staff are not horticulturally trained, and so any advice they offer may not be authoritative. The expert staff are often on duty at weekends when the stores are busiest, so seek them out if you need advice. Lastly, garden centres usually sell plants from a number of different suppliers, so the quality may vary.

Specialist nurseries usually raise their plants on-site, and their staff can offer growing advice and help you to identify the best cultivars for the conditions in your garden. Many nurseries offer mail-order services, which enable you to buy a wider range of plants. The RHS *Plant Finder* lists nurseries that supply the plants listed in the book by mail order. Plants are also available over the internet, but remember to check that suppliers' websites are secure and their business is trustworthy.

A good plant is a healthy specimen that will grow quickly and strongly. Look for new stems or buds: a multi-stemmed clematis like the one on the left is better than a single-stemmed one.

Look under the pot; if there are no or few roots growing through the drainage holes the plant is probably not root-bound.

Slide the plant out of the pot to check that the roots are pale, plump, and not congested, and that the compost is moist.

Reject a weedy pot; weeds are a sign of neglect. They stress the plant as they compete for food and water.

CHOOSING YOUR PLANTS

Sometimes it is tempting to buy plants in flower because you can see what the flowers look like, but root growth ceases while plants are blooming, so they may take longer to establish. If you are thinking of buying large plants for instant impact in your garden, remember that mature specimens are slower to recover from the shock of planting than smaller, younger ones. Large plants also have more leaves from which to lose water, and tend to dry out before their roots can establish and take up moisture. Young plants planted at the same time often overtake larger ones after a year or so.

HEALTH CHECKS

Make sure that you buy a sturdy, healthy specimen with a balanced shape and multiple stems. Leaves should also be free from signs of pests or diseases. Most plants at garden centres and nurseries are container-grown. Check the root ball; if there is a solid mat of roots, or if they coil around the base or sides of the pot, the plant has been in it for too long and may suffer from a lack of nutrients.

You usually get what you pay for with plants, so go to a well-run nursery or garden centre for healthy, top-quality plants. However, even in the best establishments, it's still worth checking the plants over carefully before you buy. Reject any plant that looks sickly or shows signs of disease or pest infestation.

WHEN TO PLANT

Although it is usually best to plant your purchases straight away, there are times when this is not possible. In these cases, care for the plants until you can plant them. Container-grown plants can be kept in their pots for a few weeks before planting out, provided you keep them well watered during dry periods.

If you buy tender bedding plants in early spring, before the threat of frost has passed, keep them in a frost-free place, such as a greenhouse or cool conservatory, or on a windowsill, until warmer weather arrives. Make sure you keep their soil moist and deadhead them regularly. In a frost-free greenhouse, you can plant bedding in hanging baskets or larger containers and grow them on so they fill out before being set outside for the summer.

PERENNIALS, ANNUALS & BIENNIALS

These plants are the mainstay of the traditional flower garden, and will probably provide you with most of the flowers in your beds and borders from early spring right through to the end of autumn. Perennials are long-lived and, although they may die down in autumn, provided they are hardy enough to survive the winter, they will come up again and produce new growth the following spring. They may well be with you for many years so, unless you're prepared to move them, give some thought to where best to site them before planting. Annuals, in contrast, only last for a year and either self-seed or are sown or planted afresh each spring. Biennials have a two-year cycle, flowering only in their second year.

A mixed border needs a little planning. Think ahead about where to place your plants according to how large they grow and when they bloom. Use annuals to fill spaces and to take over from early-flowering bulbs.

GROWING PERENNIALS

Most hardy herbaceous plants do best if planted in autumn, when the soil is still warm and in time for the winter rains to help new roots grow. Tender types prefer spring planting. There are times when you should avoid planting, such as summer or winter – new plants suffer in long spells of dry, very wet, or cold weather. Strong wind, fierce sun, and frost can all damage exposed roots. Also, avoid planting soon after heavy rain or a period of waterlogging, and don't dig or walk on clay soils when they're wet, since this can cause compaction, and squeezes out oxygen needed by plant roots.

PREPARING THE PLANTING SITE

Weeds abhor a vacuum, and any clear soil is likely to be colonized by them. If left to grow, they will dramatically reduce the amount of water, nutrients, and light that your new plants receive, just when they need it most. Take some time to clear weeds properly before planting. Use a fork to lever deep-rooted perennial weeds out of the soil; if you leave any roots, they may reshoot.

Some ornamental plants, particularly those that form dense ground cover, such as geraniums, hog the border in much the same way as weeds. Make sure that you place new plants a decent distance away from them, or cut back the ground cover until the newcomers are established. Large hedges and trees may also compete with new plants for water and light. Additionally, avoid planting closer to walls than about 45cm (18in), because the ground at the foot of a wall can be extremely dry – this area is called a rain shadow.

Delphiniums are classic herbaceous border plants. Some varieties grow very tall and tend to become top heavy, especially after rain, so they will probably need supporting.

Weeding thoroughly is essential before planting up a border. Neither new annual seedlings nor re-emerging perennials will thrive if weeds are competing with them for light, water, and nutrients.

PLANTING A PERENNIAL

Prepare the soil of the planting area to get your plant off to a flying start (*see pp.82–83*). Add organic matter to improve the soil's structure and encourage new root growth – the farther the roots spread, the greater the area from which they can obtain water and nutrients. Water the plant well about an hour or two before planting, since a dry rootball is difficult to wet when planted.

1 Stand the plant in its pot in position in the border. Do a visual check – will it have enough room to spread to its mature height and width or will it be overwhelmed by neighbouring plants?

2 Dig in plenty of organic matter, such as garden compost or well-rotted farmyard manure, over and beyond the planting area. Dig a planting hole about twice as wide and as deep as the pot or rootball.

3 Knock the plant out of its pot and gently tease some roots free from the rootball – this encourages them to spread into the surrounding soil, especially if the roots are congested. If you see any dead or damaged roots, trim them off with secateurs.

4 Stand the plant in the hole. Ensure that the top of the rootball is level with the soil surface; if needed, remove or add soil to get the level right. Draw soil around the rootball with your fingers to fill in the hole, making sure that the level of the plant doesn't change.

5 Use your heel to firm the soil gently around the plant. This removes any air pockets and makes sure that the roots are in contact with the soil. Double-check that the plant hasn't become lopsided.

6 Water thoroughly after planting to settle the soil around the roots and eliminate any air pockets. Apply a thick mulch to help keep in the moisture, and water your new plant in dry periods.

SUPPORTING PERENNIALS

There is a misconception that herbaceous plants must be both sheltered and staked. This is not always the case: bear in mind that gardens full of movement are more dynamic and interesting – while woody plants will not bend much in the breeze, flowers and grasses will billow attractively. Many tall plants are self-supporting, including foxgloves (*Digitalis*), mulleins (*Verbascum*), and *Verbena bonariensis*. Choose carefully, and you can grow a range of plants even in fairly windy sites – they often develop sturdier stems, more resistant to wind damage. Plants grown in a sheltered spot are more likely to suffer in a sudden blast of wind than those that are exposed all season long.

That said, you may want to grow some herbaceous plants with delicate or weak stems, which are prone to splay out due to the weight of their flowerheads or when beaten down by strong wind or rain. In exposed areas, supports are almost always necessary; insert them in spring, when leafy growth begins, as plants that have flopped never recover their former glory. Provided the support is a little shorter than the eventual height of the plant, it will soon be hidden by the foliage (although remember that some materials actually have a charm of their own).

Choosing suitable supports

A clump of bamboo thinned annually will yield a good supply of canes to be used as supports. Cut when young, the stems will be pliable enough to bend into hoops and semicircles. If you have space, you could also grow a hazel or willow to provide a supply of twiggy stems and thicker stakes for building rustic tripods.

Canes and pea sticks will be adequate for many perennials. You can lift plants that have flopped with canes and string, but if you bunch the stems tightly they could look worse than before. For large-flowered, tall plants, such as dahlias, chrysanthemums, and peonies, supports that allow the stems to grow through them are more effective. Metal grids on long spikes can be raised as the plants grow. For very tall dahlias and delphiniums, tie each stem to a sturdy single stake using garden string or raffia.

Where a border flanks a path or lawn, hold plants back with metal or wooden-log edging, or make your own from freshly cut bamboo canes bent into arcs. Several of these can be used to form a low temporary fence or to keep bushy growth in check, preventing it swamping lower plants.

Twiggy sticks provide a subtle and inexpensive way to support growing plants.

Long-lasting metal stakes can be linked together to encircle clumps of different sizes. Dark-coloured supports will be the least visible.

Prevent delphiniums from flopping by staking them when they are young. To avoid eye injuries, use cane tops or corks to cover the ends of canes.

Cosmos will bloom almost continuously throughout summer and up to the first frosts if you continue to deadhead it. Take the long flower stems back to the ferny foliage below.

Shearing catmint back after it has flowered in summer will stimulate the plant into producing more fresh foliage and a second, smaller flush of blooms.

DEADHEADING PERENNIALS

Picking off the dead flowerheads not only tidies up a plant, but also channels its energy into growth and producing new flowers, rather than into seed production. As a result, deadheading will prolong the flowering display of some plants, such as sweet peas (*Lathyrus*), pelargoniums, and verbenas.

What to deadhead

Although not an exact science, there are a few guidelines:
• **Large flowers,** such as those of some pelargoniums, can be snapped off individually. The soft stems of most perennials are easily pinched through.
• **If the tall flowering stems** of delphiniums, lupins, and foxgloves (*Digitalis*) are cut off, new smaller heads might sprout lower down.
• **Dainty plants** with small flowers such as lobelias are best trimmed over using scissors.
• **For repeat-flowering roses,** cut off blooms with secateurs at the cluster point, either one by one, or just above a leaf.

Shearing back

An easy way to deadhead some plants is to trim them back with shears. Trim after flowering for more flowers, fresh foliage, or both. Try this with border campanulas, catmint (*Nepeta*), hardy geraniums, knapweeds (*Centaurea*), border salvias, and pulmonarias.

In early spring, shear back growth of winter-flowering heathers (*Erica carnea*), ling (*Calluna*), periwinkles (*Vinca*), St John's wort (*Hypericum* x *hidcoteense*), and ornamental grasses, taking care not to cut into the new shoots. Also shear off old leaves on epimediums in late winter, before they flower.

LEAVE TO SEED

Not all plants need deadheading. Some will not produce more flowers, and may develop seedheads if flowers are left in place. These may be very attractive, especially in winter, and the scattered seed could result in a crop of seedlings the next year. Try leaving the flowerheads on the following plants:

- Achillea
- Astilbe
- Clematis like C. tangutica, C. orientalis, and their hybrids
- Eryngium (*below*)
- Ornamental grasses
- Poppies
- Sedum
- Teasel

PERENNIALS FOR WINTER AND SPRING INTEREST

1 *Brunnera macrophylla* 'Dawson's White' **Brunnera**
The white in the name refers to the leaf edges. The flowers, appearing in spring, are bright blue.
H 45cm (18in); S 60cm (24in).

2 *Anemone blanda* var. *rosea* 'Radar' **Anemone**
Low-growing, clump-forming anemone; in spring, has magenta-coloured flowers with white centres.
H 15cm (6in); S 15cm (6in).

3 *Arum italicum* subsp. *italicum* 'Marmoratum' **Lords and ladies**
Attractive marbled, green leaves are welcome in winter; followed by flowers in summer, and orange-red berries through to autumn.
H 30cm (12in); S 15cm (6in).

4 *Hepatica nobilis* var. *japonica* **Hepatica**
Tiny, low-growing plant whose white, pink or blue flowers open as early as late winter or early spring.
H 8cm (3in); S 15cm (6in).

5 *Bergenia* 'Sunningdale' **Elephant's ears**
Early-flowering evergreen that produces clusters of bright lilac-magenta flowers from early spring onwards.
H 30–45cm (12–18in); S 45–60cm (18–24in).

6 *Helleborus torquatus* 'Dido' **Hellebore**
'Dido' has slightly ragged, purple and lime-green flowers and narrow, deciduous leaves.
H 40cm (16in); S 45cm (18in).

7 *Primula veris* **Cowslip**
Delicate clusters of small, scented, yellow flowers held on single, upright stems.
H 25cm (10in); S 25cm (10in).

8 *Aubrieta deltoidea* **Aubretia**
Often seen cascading over walls and rock gardens, aubretia is one of the earliest and most colourful spring-flowering perennials.
H 5cm (2in); S 60cm (24in) or more.

PERENNIALS FOR AUTUMN INTEREST

1 *Catananche caerulea* 'Bicolor' **White cupid's dart**
Daisy-like, white flowers with purple centres appear
in midsummer and last well into autumn.
H 50–90cm (20–36in); S 30cm (12in).

2 *Ligularia dentata* 'Desdemona' **Ligularia**
Large, daisy-like orange flowers and heart-shaped,
leathery leaves, brownish green on top and maroon-
purple underneath.
H 1.2m (4ft); S 60cm (24in).

3 *Inula hookeri* **Inula**
Flat, yellow flowers with delicate, narrow petals and
dark yellow centres. Flowers into mid-autumn.
H 75cm (30in); S 45cm (18in).

4 *Kniphofia triangularis* **Red-hot poker** or **torch lily**
The flower spikes are shorter than other varieties,
but appear later and last into autumn.
H 1.2m (4ft); S 60cm (24in).

5 *Phlomis russeliana* **Phlomis**
Yellow flowers from late spring to early autumn;
they form pretty dried seedheads in winter.
H 1m (3ft); S 60cm (24in).

6 *Liriope muscari* **Lilyturf**
Spikes of lavender or violet-blue flowers from
early to late autumn, and long, dark-green leaves.
H 30cm (12in); S 45cm (18in).

7 *Physalis alkekengi* **Chinese** or **Japanese lantern**
Grown less for its summer flowers than for the
papery red lanterns enclosing orange-scarlet berries
that appear in autumn.
H 60–75cm (24–30in); S 90cm (36in).

8 *Salvia patens* 'Cambridge Blue' **Salvia**
Intensely blue, hooded flowers are sparse but last
from midsummer to mid-autumn.
H 45–60cm (18–24in); S 45cm (18in).

PERENNIALS FOR SHADE

1 *Aquilegia formosa* **Columbine**
Delicate, red, orange and yellow flowers.
Happy both in full sun or in full shade.
H 60–90cm (24–36in); S 45cm (18in).

2 *Ophiopogon planiscapus* 'Nigrescens' **Lilyturf**
Low-growing evergreen with grass-like, purple-black
leaves, lilac flowers in summer, and blue-black berries
in autumn.
H 23cm (9in); S 30cm (12in).

3 *Hosta* 'Francee' **Hosta**
Large, heart-shaped, olive-green leaves with
irregular white edges. Lavender-blue flowers
in summer.
H 55cm (22in); S 1m (3ft).

4 *Epimedium grandiflorum* 'Crimson Beauty'
Barrenwort or **bishop's mitre**
Ground cover plant for a shady border. Copper-
tinted green leaves and crimson flowers.
H 30cm (12in); S 30cm (12in).

5 *Pulmonaria* 'Mawson's Blue' **Lungwort**
Spreading, ground-cover plant with dark green
leaves and deep blue flowers in spring.
H to 35cm (14in); S to 45cm (18in).

6 *Pachysandra terminalis* **Pachysandra**
Spreading, shade-tolerant perennial with glossy,
evergreen leaves and small white flower spikes
in early summer.
H 20cm (8in); S indefinite.

7 *Aruncus dioicus* 'Kneiffii' **Goatsbeard**
Tiny, creamy white flower plumes stand up above
delicate, fern-like, green leaves.
H 1.2m (4ft); S 45cm (18in).

8 *Bergenia* 'Morgenröte' **Elephant's ears**
Leathery-leaved, clump-forming evergreen with
deep carmine flowers; thrives in partial shade
beneath trees.
H 30–45cm (12–18in); S 45cm (18in).

PERENNIALS FOR CUT FLOWERS

1 *Alstroemeria ligtu* hybrids **Peruvian lily**
Funnel-shaped, lily-like flowers in a wide range of
colours – from pinks and reds to yellows and oranges.
H 45–60cm (18–24in); S 0.6–1m (2–3ft).

2 *Aster novi-belgii* 'Lady in Blue' **Michaelmas daisy**
Masses of daisy-like flowers produced from late
summer to mid-autumn.
H 30cm (12in); S 50cm (20in).

3 *Paeonia officinalis* 'Rubra Plena' **Peony**
Large, double, vivid pink-crimson flowers with
satiny, ruffled petals in early to midsummer.
H 75cm (30in); S 75cm (30in).

4 *Dahlia* 'David Howard' **Dahlia**
Miniature variety with rounded, bronze-orange
flowers and dark-bronze foliage.
H 90cm (30in); S 60cm (24in).

5 *Delphinium* 'Fenella' **Delphinium**
A traditional, old variety producing tall spikes
of purple-tinged, gentian-blue flowers with
black eyes.
H to 1.5m (5ft); S 75cm (30in).

6 *Helenium* 'Crimson Beauty' **Helen's flower**
Daisy-like flowers with drooping petals in deep
copper-red that turn brownish as they age and
with prominent, domed centres.
H 90cm (36in); S 60cm (24in).

7 *Lychnis chalcedonica* **Jerusalem** or **Maltese cross**
Flat heads of small, cross-shaped flowers that are a
brilliant vermilion in colour in early and midsummer.
H 1–1.2m (3–4ft); S 30–45cm (12–18in).

8 *Gaillardia* x *grandiflora* 'Dazzler' **Blanket flower**
Large daisy-like flowers, up to 15cm (6in) across with
a domed orange-red centre and bright orange-red
petals with yellow tips.
H 60cm (24in); S 50cm (26in).

GROWING ANNUALS AND BIENNIALS

You can achieve fantastic results quickly by sowing hardy annuals directly into the soil in spring, either to create an entire annual display or to fill gaps in beds and borders. It works out much cheaper than buying summer bedding in trays. Three weeks before sowing, weed and fork over the area; two weeks later, clear any weeds. This reduces the amount of weeding you need to do later, as well as eliminating competition for the water and nutrients that seedlings need. Biennials can be treated in much the same way as annuals; because they are left in the ground the winter after sowing, though, they may need protecting from frost.

SOWING FOR SUCCESS

Annuals and biennials look more natural if sown in informal swathes, or drifts, that merge into one another. To obtain this effect in a bed or border, draw out irregularly-shaped areas, one for each annual, in the soil with a cane. Then mark out straight rows (*see p.83*) within the drifts and sow one type of seed in each. Sowing in rows will make it easier to distinguish the annual and biennial seedlings from weed seedlings. Then you can easily remove weeds by hoeing between the rows.

Once you have sown the seeds, don't forget to label them. In dry weather, water the seedlings using a watering can with a rose – a spout causes runnels in the soil and puddling, which can wash seedlings out of the ground, and in some soils leaves a hard crust on the surface.

Half-hardy annuals and biennials usually need to be sown in containers so they can develop undercover (*see pp.300–303*); they can be planted out once the weather warms up. Check the seed packet for more details. Remember that biennials will not flower until their second year; some will need protecting from frost during winter (known as overwintering) using cloches or horticultural fleece.

Hardy annuals provide an easy way to introduce a splash of colour into your garden. Here, orange pot marigolds add zing to the soft blues and purples of love-in-a-mist and cornflowers.

COLLECTING SEEDS

Annuals produce seedheads in a variety of shapes, such as pods or papery capsules, and most are easy to collect for sowing early the following spring. It is not worth collecting seed from F1 hybrids – the seedlings won't look like the parent plant.

On a warm, dry day, when the seedheads have turned brown (*below*, love-in-a-mist), snip them off and lay on sheets of kitchen- or newspaper to dry.

When they are completely dry, shake the seeds out onto a piece of paper. Pick out any chaff, and store the seed in a labelled paper bag or envelope (plastic bags encourage mould). Store them in a cool, dry place out of bright light.

HOW TO SOW ANNUALS AND BIENNIALS

Before sowing, prepare the soil well, removing all the weeds, and forking it over. Most annuals and biennials like free-draining soil, so if you have heavy clay, dig in some coarse sand or grit to open it up. Choose an open, sunny site, not too close to large trees or shrubs which will reduce germination rates.

Always follow the recommendations on the seed packet for correct planting depths and spacing. Sow as thinly as possible, and once the seeds have germinated and the seedlings have produced a few leaves, thin them out by removing all but the strongest, leaving the correct space between each young plant.

1 Rake the soil to level it, and clear away any large stones. Use the head of the rake to break down lumps to create a crumbly soil.

2 Use a bamboo cane to create some straight drills in which to sow the seed – press the cane lightly into the soil, as shown.

3 Transfer a little seed to the palm of your hand, and gently tap it with your finger to trickle the seed gradually into the drill.

4 Cover the seeds by gently raking a fine layer of soil over the drills. Then firm the soil lightly with the head of your rake.

5 Water well, using a watering can with a fine rose. A sprinkling is less likely to disturb the seeds or cause a soil crust to form.

6 Protect seeds from being scratched up by birds and cats by covering the area with some twiggy sticks or a layer of netting.

EASY ANNUALS FROM SEED

1 *Papaver somniferum* **Opium poppy**
Pink, red, mauve, or even white flowers give way to seedheads that can be either cut for dried arrangements or left to self-seed.
H to 1.2m (4ft); S to 30cm (12in).

2 *Nigella damascena* **Love-in-a-mist**
Good as long-lasting cut flowers; the seed capsules can be dried for flower arrangements. Self-seeds freely.
H to 50cm (20in); S to 23cm (9in).

3 *Lathyrus odoratus* **Sweet pea**
A wide range of flowering annuals, most are climbers and many are highly scented. Flowers in almost every colour except yellow.
H 2–2.5m (6–8ft); S to 1m (3ft).

4 *Eschscholzia californica* **California poppy**
Fragile, poppy-like flowers open fully only in sun and vary in colour from yellow, orange and gold to red, pink and white. Almost indestructible.
H to 30cm (12in); S to 15cm (6in).

5 *Helianthus* 'Triomphe de Gand' **Sunflower**
Mid-sized sunflower with golden-yellow flowers up to 12cm (5in) in diameter.
H 1.5m (5ft); S 1.2m (4ft).

6 *Lagurus ovatus* **Hare's tail**
A compact, annual grass with pale green leaves and fluffy, hairy seedheads. Sow in groups or drifts in spring.
H 50cm (20in); S 30cm (12in).

7 *Cosmos bipinnatus* Sensation Series **Cosmos**
Flowers well into late summer, so sow in situ in
late spring to fill gaps in a border left by early
flowering plants.
H 90cm (36in); S 45cm (18in)

8 *Callistephus chinensis* Ostrich Plume Series
China aster
Sow during spring for feathery, pink, crimson or
white flowers from late summer until late autumn.
H to 60cm (24in); S 30cm (12in).

9 *Nicotiana* 'Lime Green' **Tobacco plant**
Bright, lime-green flowers have a wonderful evening
and night-time fragrance. Flowers from midsummer
to autumn.
H 60cm (24in); S 25cm (10in).

10 *Calendula officinalis* **Pot marigold**
Daisy-like flowers range in colour from bright yellow
and orange to cream and apricot. Will continue
flowering until the first autumn frost.
H 30–75cm (12–30in); S 30–45cm (12–18in)

11 *Tropaeolum majus* **Nasturtium**
Some climb, while others scramble. All grow
vigorously and produce masses of red, orange
or yellow flowers.
H 1–3m (3–10ft); S 1.5–5m (5–15ft).

12 *Ipomoea purpurea* **Common morning glory**
A climbing annual with trumpet-shaped flowers
that may be blue-purple, magenta, pink or white.
Needs full sun.
H 5.5m (18ft); S 3m (10ft).

BULBS

Regardless of the size of your garden, there is always space for bulbs. They discreetly die down when not in bloom, and their compact size means that they can easily be tucked in among other plants. When they do perform, they offer a high ratio of flower to foliage. If you plan ahead, you could have bulbs flowering in your garden nearly all year round. Bulbs can be left to naturalize in lawns or to carpet shaded woodland areas in spring. They also provide the early season's show in your borders, before herbaceous plants have stirred into growth, and make excellent companions for perennials and deciduous shrubs, filling the gaps between these plants. This chapter covers not only true bulbs but also corms, rhizomes, and tubers.

A mixed planting of early-flowering, bright red 'Showwinner' tulips and white and purple crocuses creates a startling blaze of spring colour.

GROWING BULBS

Bulbs are easy to grow, as long as you follow a few simple rules. In the wild, most bulbs grow in climates with hot, dry seasons, and prefer light, free-draining soils in full sun. Bulbs from woodlands need moist, rich, but free-draining soil, and dappled sun or part shade. In short, if you know where your bulbs come from, it's easier to provide a home for them. For example, woodland bulbs, such as bluebells, colchicums, cyclamen, and snowdrops, come into growth when the weather is sunny and wet – this will be when overhanging trees are leafless, since the foliage reduces the amount of light and rainwater reaching the bulbs. The bulb-flowering season can be quite long with careful planning, from familiar spring flowers such as daffodils through to autumn crocuses.

WHAT IS A BULB?

It can be confusing when you discover that the word "bulb" is often used as a blanket term for any plant that has an underground food-storage organ. So as well as true bulbs (as in snowdrops and lilies), there are three other forms: corms (crocuses), rhizomes (irises) and tubers (cyclamen). Each type of bulb has evolved to cope with distinct growing conditions.

True bulbs are made of leaves layered as in an onion, with a papery skin. Lily bulbs are looser in form, and lack a papery skin.
Corms are the thickened base of the stem; they have 1–2 buds at their apex. Tiny cormlets may grow around the base.
Rhizomes are swollen stems, which usually grow horizontally near the soil surface, and have several buds along them.
Tubers come in two types. Root tubers (as in dahlias) have buds at the stem base. Stem tubers (as in cyclamen) have surface buds.

IDENTIFYING BULBS

Getting to grips with what the different types of bulbs look like will help you to determine their planting needs.

clockwise, from top left:
True bulb (e.g. daffodils, tulips, and lilies)
Tuber (e.g. dahlias and cyclamen)
Rhizome (e.g. irises and lily-of-the-valley)
Corm (e.g. gladioli and crocuses)

(clockwise from top left)
Daffodils, dahlias, irises, and gladioli

A succession of flowers from bulbs can be enjoyed through much of the year, not just in spring. Try starting with snowdrops in late winter, followed by lilies in summer and nerines in autumn.

PLANNING A YEAR-ROUND DISPLAY

The late winter and early spring show is dominated by bulbs with flowers in clear yellows and whites, such as trilliums, winter aconites, and snowdrops. These are soon followed by the first daffodils, like *Narcissus* 'February Gold'. The pretty, scented pheasant's eye daffodils are among the last of the season to bloom. Early irises, hyacinths, grape hyacinths, and ipheions belong to the blue brigade that cheers spring gardens. Crocuses, of course, encompass all three colours.

It is not until the tulips emerge that the colour spectrum broadens. Tulips come in every colour except blue, and range from the dainty (*Tulipa sylvestris*, great for naturalizing) to the seriously baroque. The parrot tulip 'Rococo' has fiery red, fringed petals with green streaks.

As spring turns to summer, alliums (ornamental onions) decorate our borders. Their distinctive flowerheads, in globes or pendent clusters, range from the cheerful golden *Allium moly* and tall 'Purple Sensation' to the giant spheres of metallic, starry flowers of *Allium cristophii*, which are great for drying. More irises add dusky hues of blue, amber, rust, and purple.

Fewer bulbs perform in midsummer, but they include some stunners. Lilies are elegant and stately; many have strong perfumes. Gladioli add height. Eucomis, with its flowerheads of greenish or reddish, starry flowers topped by a pineapple-like tuft, makes a wonderful architectural plant for late summer. Finally, in the autumn there are crocus-like colchicums and vivid pink nerines.

CHOOSING AND BUYING BULBS

Most bulbs are available for purchase in their dormant state, so buy spring-flowering bulbs in late summer or autumn and summer-flowering ones in spring. Don't leave it too late; if you buy bulbs soon after they come into stock, you'll get the pick of the healthiest and most vigorous.

Choose bulbs that are big, firm and plump; plants from dried-up bulbs will not thrive. Reject any with blemishes – indications of pest damage or disease – or signs of rot or mould, or any that are missing their papery skins or have started sprouting. Look for bulbs that are labelled "from cultivated stock" or "nursery propagated" to avoid buying any that are collected from the wild.

PLANTING BULBS

Here are a few tips you should follow when planting to get the most out of your bulbs:
• **Most bulbs** should be planted 2–3 times the depth of the bulb; the exception is for tulips, which should be planted 3–4 times the depth.
• **Space bulbs** at distances 2–3 times their width.
• Plant corms and tubers with their growth buds upwards – they are not as distinct as on true bulbs.
• **Plant scaly bulbs,** such as lilies, on their sides to stop them accumulating moisture in between their scales and rotting.
• **If your soil** is cold, wet, or poorly drained, grow bulbs in pots.
• **If your soil** is light and dry, add well-rotted organic matter to prevent the roots drying out.

Bulbs in mixed borders

When planting spring bulbs in borders, you may be tempted to plant them close to the front because they flower early in the season. However, since most other plants in the border won't be very visible, you can spread bulbs around in little clusters or drifts, from the front to the back. Then, as their foliage dies down, they will be hidden behind the emerging growth of shrubs or perennials. Later-flowering bulbs are often better scattered in between other plants which also help to disguise their unsightly dying foliage.

PLANTING BULBS IN A BORDER

Most bulbs prefer free-draining soil as they like to be fairly dry when dormant; fork some grit into the planting area on heavy soils. Bulb planters take out a plug of soil for each bulb, but can be laborious on stony soil or over large areas. Instead, dig out an entire planting area.

1 Position the bulbs in the hole with their pointed tips upwards. Discard any bulbs that show signs of disease or rotting.

2 Fill in the hole with soil, firming as you go with your fingers, taking care not to damage the bulbs' growing tips.

3 Chicken wire over the planted area will prevent animals from digging them up. Remove it as soon as the first shoots appear.

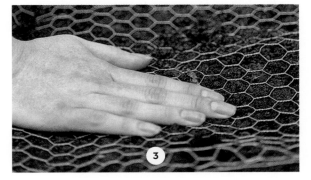

PLANTING BULBS IN GRASS

It you want to create a natural-looking display of bulbs in a lawn, make life easy for yourself by adapting your planting technique to the size of the bulbs.

To plant small bulbs, cut an H-shaped cut in the turf and fold back the flaps (*top*). If necessary, dig out the soil to the recommended depth for the bulb type. Place bulbs randomly, growing point upwards. Replace the soil, then fold back the turf and firm.

To plant large bulbs in grass, start by scattering the bulbs randomly over the ground. Use a bulb planter to make a hole, then place the bulb inside (*bottom*). Fill in the hole with soil from the bottom of the core, then replace the circle of turf, firming in gently.

(top) **Small bulbs** should be planted in the soil with the pointed growing end facing upwards.

(bottom) **To plant larger bulbs,** press the plant cutter firmly into the soil to make a neat hole, then drop the bulb in pointed end up.

Planting "in the green"

Winter aconites, snowdrops, and anemones have tiny bulbs that dehydrate very quickly and often fail to produce decent plants. If you buy these bulbs, soak them overnight in water before planting to help kickstart them into life. It is best, though, to buy plants "in the green", which means they have been lifted directly after flowering while the foliage is still fresh. This gives them a chance to re-establish before becoming dormant. Although plants in the green are more expensive, they are more reliable. If you have clumps in the garden, lift the bulbs as soon as they have flowered, tease them apart, and replant where you want them. Make sure that the pale bases of the leaves are below the soil.

Bulb aftercare

While they are in growth, bulbs need feeding, watering, and deadheading. Tall-stemmed bulbs may also need staking. After flowering, most bulb foliage dies back. Some gardeners bend the foliage over and tie it up to make it look tidy, but this stops nutrients from the leaves getting to the bulb and helping to form flower buds for next year. Let the leaves die down naturally until they are dry and yellow (about five weeks), and then cut them off.

To keep bulbs flowering well, divide them every 3–4 years. Using a fork, lever out a dormant clump, tease apart, discard any diseased bulbs, and replant. Tulips may stop blooming if they're not lifted after flowering each year. After lifting, clean the bulbs, store in paper bags in a cool, dry place, then replant them in autumn. Dahlias, gladioli, and tuberous begonias should be lifted and stored over winter.

LIFTING AND STORING IN WINTER

Except in mild areas where frost is light or non-existent, lift tender bulbs and tubers over winter. They can be kept in any cool, frost-free shed, greenhouse, or conservatory, maintaining a minimum temperature of 4–7°C (40–45°F). Cut down the stems of begonias and lift the tubers just before the frosts start. Dry them off and dust with sulphur before packing them into trays of dry compost or bark.

Lift gladioli in mid-autumn when the leaves are beginning to turn yellow. Snap off the stems and dry off the bulb-like corms. You'll find there is a withered old corm under the new one; pull this off and discard it. You will also find baby corms, which you can grow on like seeds. Dust the new corms with sulphur dust and store them in shallow trays either in a cool, dry, airy shed or in a room indoors.

Chocolate cosmos (*Cosmos atrosanguineus*) should be lifted in autumn as a precaution in very cold areas. Dig up tubers before the frosts start, and store under cover in trays of barely moist compost. Check all stored bulbs and tubers regularly for signs of rot throughout the winter, and discard any that have succumbed.

Lift bulbs with a fork once the plant's leaves turn yellow. Clean off the soil, remove the foliage, and leave them indoors to dry overnight.

STORING DAHLIAS OVER WINTER

1 Cut old stems back to a height of about 15cm (6in) above the ground. Fork the soil gently to loosen it, and carefully lift out the tubers.

2 Store tubers upside-down somewhere frost-free until they have dried out completely. Then up-end them, cover them with bark or vermiculite, and keep them dry until spring.

3 Plant out tubers in spring and cover with soil so the crown at the base of the stem, where new shoots will appear, is about 2.5–5cm (1–2in) below ground.

BULBS FOR SPRING

1 *Gladiolus tristis* **Gladiolus**
Aromatic, pale yellow or cream flowers on thin, arching stems in spring. Gladioli grow from corms.
H 45cm–1.5m (18in–5ft); S 5cm (2in).

2 *Iris* 'George' **Iris**
A vigorous but short variety that grows from bulbs. Velvety, rich purple flowers.
H 13cm (5in).

3 *Narcissus* 'Suzy' **Daffodil**
A mid-spring daffodil with large, open flowers in primrose yellow with bright orange centres.
H 40cm (16in).

4 *Chionodoxa* 'Pink Giant' **Glory of the snow**
Not 'giant' at all but a small, low-growing bulb with delicate, pink flowers.
H 10–20cm (4–8in); S 5cm (2in).

5 *Muscari armeniacum* **Grape hyacinth**
Spikes of tightly clustered, tiny, blue flowers in spring. Perfect for containers and rock or woodland gardens.
H 40cm (16in); S 5cm (2in).

6 *Crocus minimus* **Crocus**
A deep lilac- or purple-coloured crocus that does not flower until late spring.
H 8cm (3in); S 2.5cm (1in).

7 *Hyacinthus orientalis* 'Ostara' **Hyacinth**
Large, upright spikes of violet blue flowers. Good for borders or containers.
H 20–30cm (8–12in); S 8cm (3in).

8 *Fritillaria meleagris* **Snake's head fritillary**
Nodding, bell-like, purple and white flowers, excellent for naturalizing in grass or meadow gardens.
H to 30cm (12in); S 5–8cm (2–3in).

9 *Anemone pavonina* **Anemone**
A Mediterranean variety that likes full sun and blooms
in spring. Flowers may be red, pink, or purple.
H 25cm (10in); S 15cm (10in).

10 *Galanthus* 'Atkinsii' **Snowdrop**
Single, nodding, white flowers appear as early as
late winter and into early spring.
H 20cm (8in); S 8cm (3in).

11 *Narcissus* 'Satin Pink' **Daffodil**
An unusual variety with pure white petals and a
pink-flushed central cup. Flowers in mid-spring.
H 45cm (18in).

12 *Eranthis hyemalis* **Winter aconite**
Bright yellow flowers in late winter and early spring.
Aconites grow from small, knobbly tubers.
H 5–8cm (2–3in); S 5cm (2in).

13 *Erythronium* 'Pagoda' **Dog's-tooth violet**
Sulphur-yellow flowers in spring, backed by glossy,
deep green leaves with bronze mottling. Suited
to moist, shady woodland.
H 15–35cm (6–14in); S 10cm (4in).

14 *Tulipa* 'Ballerina' **Tulip**
Sweetly scented, flame-coloured blooms in
blood-red and orange-yellow.
H 60cm (24in).

15 *Ornithogalum dubium* **Star-of-Bethlehem**
Unusual species with waxy, yellow-orange flowers.
H 15–30cm (6–12in); S 15cm (6in).

16 *Camassia leichtlinii* subsp. *leichtlinii* **Quamash**
Tall spikes of star-shaped, creamy white flowers.
Good for cutting.
H 60cm–1.2m (24in–4ft); S 10cm (4in).

BULB-LIKE PLANTS FOR SUMMER AND AUTUMN

1 *Dahlia* 'Arabian Night' **Dahlia**
Large, double, rich burgundy-red flowers from early summer, from tubers.
H 1.2m (4ft); S 50cm (20in).

2 *Iris latifolia* **English iris**
Mid-sized, narrow-leaved iris with flowers in blue, violet, or white.
H 25–60cm (10–24in).

3 *Lilium henryi* **Lily**
Flowers in late summer, with deep-orange blooms spotted with black markings.
H 1–3m (3–10ft).

4 *Roscoea cautleyoides* **Roscoea**
Useful as a plant that likes damp, shady sites and flowers all summer and into autumn. It grows from tubers.
H 55cm (22in); S 15cm (6in).

5 *Lilium regale* **Regal lily**
Fragrant, white, trumpet-shaped flowers with pink-veined petals and yellow centres appear in summer.
H 1–1.5m (2–5ft)

6 *Allium cernuum* **Nodding onion**
Stiff, upright stems support small, drooping, deep pink, bell-shaped flowers.
H 30–60cm (12–24in); S 5cm (2in).

7 *Colchicum speciosum* 'Album' **Autumn crocus**
Small white flowers appear in autumn before the leaves. Corms can be planted in grass.
H 18cm (7in); S 10cm (4in).

8 *Tigridia pavonia* **Tiger flower**
Short-lived exotic needing full sun. Iris-like flowers in summer. Lift bulbs in winter.
H 1.5m (5ft); S 10cm (4in).

9 *Crocus ochroleucus* **Autumn crocus**
Tiny, late-summer-flowering crocus with creamy-white flowers with yellow throats.
H 5cm (2in); S 2.5cm (1in).

10 *Fritillaria camschatcensis* **Black sarana** or **chocolate lily**
The dark purple-black flowers appear in early summer, but beware: they have a notoriously unpleasant odour.
H to 45cm (18in); S 8–10cm (3–4in).

11 *Begonia* 'Herzog von Sagan' **Begonia**
Large, yellow flowers with pink, frilled margins. Lift the tubers in autumn before frosts.
H to 90cm (36in); S 45cm (18in).

12 *Galtonia viridiflora* **Galtonia**
Arching stems support nodding, green-and-white flowers in late summer.
H to 1m (3ft); S 10cm (4in).

13 *Nerine bowdenii* **Nerine**
Small but flamboyant, lily-like flowers in bright pink appear in autumn, when most other colourful blooms are fading.
H 45cm (18in); S 8cm (3in).

14 *Scilla peruviana* 'Alba' **Squill**
Tight, conical clusters of small, star-shaped, white flowers in early summer.
H 15–30cm (6–12in); S 5–10cm (2–4in).

15 *Canna* 'Endeavour' **Indian shot plant**
Tall stems of bright red, iris-like flowers from midsummer to early autumn. Cannas grow from rhizomes.
H 1.5–2.2m (5–7ft); S 50cm (20in).

16 *Cyclamen hederifolium* **Cyclamen**
Traditionally grown under trees, the pink flowers appear in mid- and late autumn, before the leaves. Grows from tubers.
H 10–13cm (4–5in); S 15cm (6in).

SHRUBS

Shrubs are vital furnishings for the garden. They provide a permanent structure, adding seasonal interest. They can also be used to define areas of the garden, and to disguise eyesores, such as bins and compost heaps, as well as neighbouring gardens. Shrubs are incredibly versatile and there is one to suit every design and location, whether your garden is large or small. They range from dwarf, ground-covering plants, suitable for the front of a border, to large architectural specimens that grow to a couple of metres tall and wide, and make great screens or backdrops.

A flowering shrub, such as *Hydrangea paniculata* 'Unique', provides a double-whammy: attractive green foliage all spring and summer plus beautiful, conical, white flowerheads in August and September.

GROWING SHRUBS

Use shrubs to help define your garden style. For a modern or minimalist design, consider mass plantings of one plant, or two contrasting plants, such as the evergreen *Viburnum davidii*, with its small white flowers followed by steely blue berries, teamed with clipped golden privet, *Ligustrum ovalifolium* 'Vicaryi'. In a more traditional setting, use a wider range of plants for diversity and interest. Some shrubs, such as *Viburnum plicatum* f. *tomentosum* 'Mariesii', with its tiers of horizontal branches, have such a striking appearance that they can be used to provide a focal point at the end of a vista, or a centrepiece in a lawn or gravel garden.

SHRUBS FOR STRUCTURE

Ideal as background plants, shrubs provide solidity and structure, creating a sense of permanence in a planting scheme. Avoid planting in a uniform line against a fence; instead, use plants of different heights and shapes strategically in beds and borders. Use large, dark specimens behind light-coloured flowers; small shrubs, such as dwarf box, thyme, or euonymus to edge the front of a scheme; and those with dramatic foliage, such as *Fatsia japonica*, *Sambucus racemosa* 'Sutherland Gold', or a tree peony (*Paeonia delavayi*) to provide focal points.

Remember that many shrubs have pretty flowers and berries; seasonal features should be considered carefully when making planting plans. For example, while the white-flowered *Choisya* x *dewitteana* 'Aztec Pearl' will set off the oval blue flowerheads of *Ceanothus* 'Blue Mound' brilliantly, the orange blooms of a kerria matched with pale pink montana clematis 'Elizabeth' creates a sickly mix.

The red autumn berries of cotoneaster are a good meal for birds stocking up for winter. Site one near a window to watch them feeding. The white spring flowers also attract insects.

Nectar-rich plants, such as the butterfly bush, *Buddleja davidii*, will attract butterflies and moths into your garden with its mauve, pink, red, or white summer flowers.

YEAR-ROUND INTEREST

To create an exciting shrub border, compile a list of plants that perform at different times of the year.

• **Winter and early spring** can make gardens appear bare, so choose plants that flower at this time. The yellow flowers of forsythia and pink flowering currants (*Ribes sanguineum*) kickstart spring, together with the early magnolias. Certain shrubs that flower in winter and early spring have a wonderful perfume – such as witch hazel (*Hamamelis*), daphne, and winter box (*Sarcococca confusa*). The corkscrew hazel (*Corylus avellana* 'Contorta') sports wild, twisted stems, and dogwoods (*Cornus*) brighten up winter gardens with colourful bark. Evergreens with different leaf shapes and textures will help to increase visual contrast.

• **In spring and summer,** shrubs that flower include berberis, lilac (*Syringa*), and scented osmanthus and philadelphus; smaller shrubs make colourful fillers in the border. For summer colour, plant buddlejas, hydrangeas, escallonias, fuchsias, and hypericum. Summer-flowering heathers will thrive if planted in acid soil.

• **Autumn** will often bring a cold snap after a warm summer, producing fantastic foliage colour. Winning plants for this time of the year include Japanese maples, deciduous cotoneasters, and azaleas, dogwoods (*Cornus*), and the smokebush (*Cotinus*). Birds soon devour certain fruits and berries; to extend the season of interest, combine shrubs with a few whose fruit is less attractive to wildlife, such as some crab apples (*Malus*), yellow-berried cotoneasters, and roses.

(opposite) **The unusual shape** of *Viburnum plicatum* f. *tomentosum* 'Mariesii' makes it an eye-catching focal point in a border. This large shrub has horizontally layered branches that are laden with white flowers in spring.

CHOOSING AND BUYING SHRUBS

At the garden centre, look for healthy plants in clean, damp compost that show no sign of pests or diseases. Reject plants that look in poor health as they may be carrying a disease that will affect your other plants. Don't be tempted by the "sales" corner – the plant is usually only there because it has been standing around too long, and is unlikely to do well once planted.

• **Leaves** should not be discoloured, pale, or wilting. Look for specimens with bright, new leaves or shoots, as this shows that they are growing well.

• **Roots** should not be poking through the holes at the base or growing in circles indicating the plant is pot-bound; tap the rootball out of its pot to check. A healthy root system means that the plant should quickly establish after planting.

• **Top-heavy plants** should be avoided; their roots are likely to have become congested.

• **Dead leaves** are a sign of neglect or disease. If diseased plants and their compost are brought into the garden, the complaint may spread to other plants.

• **Weeds** growing on the compost are a sign of neglect, and will have deprived it of water and nutrients. They may also be carrying diseases that can affect the plant.

Leaves bearing holes could be a sign that the plant is infected by pests or diseases.

Check the plant's roots: if they are densely congested, the plant is likely pot-bound and should not be bought.

PLANTING A SHRUB

If you have just bought a container-grown shrub, first water it well by plunging it – pot and all – in water for about an hour, then allow it to drain. To remove the rootball from the pot, turn the pot over and hold the shrub firmly with the fingers of one hand between the stems at soil level, then gently squeeze the sides of the pot, or tap the base, until it slides off. Carefully tease out any roots that are curling around the rootball.

1 Dig a planting hole that is as deep and twice as wide as the rootball of the shrub (here a viburnum). Place the plant into the hole and lay a cane across the top to make sure that it is at the same depth as it was in its original container. Adjust the depth of the hole if necessary.

2 Set the shrub in the hole, and then backfill around the rootball with your hands, firming the soil with your heel. At the same time, create a slight saucer-shaped depression within the soil surrounding the plant – this will encourage water to collect naturally over the root area.

3 Prune lightly only if necessary: cut out any stems that are dead or unhealthy, and prune any that are growing towards the centre of the plant back to a healthy, outward-facing bud or shoot. Water the plant well, and then apply a mulch, making sure that it does not touch the stems.

TRANSPLANTING A SHRUB

If you decide that a shrub is in the wrong place, you can simply move it. The younger the shrub, the easier and more successful the move should be. The best time to transplant a deciduous shrub is in the autumn; move evergreens in the spring, just before the new growth emerges. Prune the shrub back by a third before transplanting – this will compensate for root loss, and should make it easier to handle.

1 Mark out a circle around the outer edge of the roots using a spade. The roots of shrubs normally do not spread out further than the extent of the stems before they were pruned.

2 Dig a trench around the circle and use a fork to loosen the soil around the roots. With a spade, dig under the rootball so that it can be lifted out of the soil, cutting through woody roots if necessary.

3 Tilt the plant to one side and feed a piece of hessian or a ground sheet under it. Then rock the shrub to the other side and pull the material through so that the plant is sitting on top of it.

4 Enlist the assistance of someone else to help you carry the shrub to its new location in the garden. This is especially important with a large, heavy rootball like this one.

5 Dig a hole that is twice as wide and the same depth as the rootball at the new site. Carefully lower the shrub into the new hole.

6 Fill in with soil around the roots, ensuring there are no air gaps. Firm the soil with your foot. Water in the plant well, and apply a mulch.

PRUNING SHRUBS

Keep an eye on your shrubs throughout the year, cutting out dead, diseased, and damaged stems to keep the plants healthy. Remove any stems that are crossing and rubbing against each other, all-green shoots on variegated plants, and suckers on grafted plants. Cutting out badly placed stems and reshaping plants will improve their appearance. Some pruning methods increase growth, fruiting, and flowering, as well as stimulating the production of young, decorative stems.

WHEN TO PRUNE

As a general rule of thumb, prune late-summer and autumn-flowering plants, which usually bloom on the current year's stems, in early spring. Spring and early-summer flowering plants bloom on the previous year's wood, and should be pruned after flowering, so that the new stems have time to ripen before winter. To renovate most deciduous trees and shrubs, prune in winter when the plants are dormant. For the majority of evergreens, prune in spring, just as the new growth resumes.

Bear in mind that pruning deciduous plants in winter results in vigorous regrowth in spring. Pruning in spring and early summer produces much thinner growth, while summer pruning reduces the leafy canopy and the plant's food resources, resulting in relatively weak regrowth.

Mophead hydrangeas (*Hydrangea macrophylla*) will flower on stems produced the previous year. In mid-spring, prune old stems by up to 30cm (12in), down to pairs of fat, healthy buds – these will produce the flowering shoots.

To encourage new stems on multi-stemmed plants such as *Viburnum x bodnantense*, cut out any old weak stems at the plant's base. If the plant is overgrown and flowering is poor, cut down all of the stems in late spring.

Light pruning

If you are not familiar with your shrubs, prune them lightly, and see how they respond during the growing season. Prune newly-planted shrubs cautiously, as they need to grow to develop their natural beauty – although you may wish to encourage a good shape with a little pruning. For example, box (*Lonicera nitida*) or yew (*Taxus*) can be clipped lightly into a sphere. Other plants, like heathers, just need shearing over to remove spent flowers, and to promote neat growth. Tip pruning, where the shoot tips are removed,can be carried out on young shrubs to encourage a bushy habit.

Many plants that should be pruned cautiously will respond to renovative pruning when they become too large for their site, their stems have become congested, or flowering is poor. In these cases, cut back the stems to near ground level when the plant is dormant, or for evergreens, buds are just breaking, in late winter or early spring.

Moderate pruning

Cutting shrubs back to half their size works for those that resent being pruned to a low framework, but flower on new wood, such as sun roses (*Cistus*). Prune shrubs that flower on shoots produced the previous year, such as broom, after flowering. Remove the old flowering stems and shorten new stems to encourage more sideshoots – they will bloom the following year.

Hard pruning

Cutting back to soil level works for plants that send up shoots from below the ground, such as brambles (*Rubus*). The resulting canes are either colourful, or will flower and fruit prolifically. Leave a low framework of stems, 20–30cm (8–12in) high, from which new shoots can arise. Plants that produce young, coloured stems, such as dogwoods (*Cornus*) and willows (*Salix*), can be pruned in this way in late winter.

Prune young shrubs, such as this mahonia, in mid-spring; this will stimulate the growth of more flowering stems which will bloom in the coming autumn/winter. Prune off the old flowering stems, cutting just above a leaf or sideshoot.

Old wood tends to be unproductive, channelling energy away from flowering young stems. Remove it with a clean cut just above the new stem. Make a slanted cut so that rainwater will not collect around the base of the new wood and rot it.

PRUNING AN EARLY-FLOWERING SHRUB

Shrubs that flower before midsummer, such as deutzia (*shown here*), forsythia, and kerria, should be pruned after they have flowered. Their blooms develop on stems that were made the year before, therefore pruning soon after flowering gives the plants time to develop new stems which will bloom the following year. Deutzias can become tall and leggy if left unpruned, but regular trims keep them bushy and compact, and encourage more stems of sweetly-scented flowers to develop.

1 Inspect your plant: this deutzia is five years old, and has become too tall, while the stems shooting up from the base are congested. Some flower buds are also suffering from frost damage.

2 Cut out any wood that is dead, diseased, and damaged. Then use sharp secateurs to remove one in three of the congested stems, targeting the old and weak growth.

3 Shorten the remaining stems to a pair of strong, healthy buds, or new shoots, as shown here. Pay particular attention to crowded shoots in the centre of the plant.

4 Remove any stems that are crossing and rubbing against each other. Also prune out shoots that are growing towards the middle of the plant.

5 Reduce the height of the shrub if necessary: take down the tallest stems, cutting back to healthy, outward-facing shoots or buds.

6 The pruned shrub is shorter and the congestion has been reduced. New growth will soon appear from the base and below pruning cuts, resulting in more flowers.

PRUNING FOR GOOD FOLIAGE

Hazel, elder, willow, and coloured-stemmed dogwoods can all withstand hard pruning. When grown for their foliage effects, as with this variegated elder, or for their red or golden shoots, as with many dogwoods and willows, all the old stems should be pruned in late winter or early spring in order to encourage new growth.

1 Elder is a vigorous plant – most of the growth seen here was made during the previous year. In alternate years, all the stems can be pruned back in early spring to a low framework near its base.

2 Prune each stem back to two or three pairs of buds above the base of the previous year's wood. Although you will get the best foliage by pruning in this way, you are also removing the flowering stems.

3 Cut old, congested wood in the centre of the plant right down to the ground. Also remove any stems that are badly positioned, or growing towards the middle.

4 Make smooth, clean cuts, using loppers on all branches that are more than 1cm (½in) in diameter. Use a curved pruning saw on very thick branches, or where you need to make a cut in an awkward spot.

5 Pruning will encourage vigorous growth, despite the finished result looking sparse. Now the flowering stems have been removed, this spring-flowering plant will not bloom or produce berries this year.

6 By early summer, the elder's stems will have grown by 1m (3ft) or more, and the fresh, beautifully variegated leaves will unfurl to make a compact, eye-catching feature in the garden.

SHRUBS FOR SPRING COLOUR

1 *Rhododendron* 'Irohayama' **Azalea**
Low-growing, evergreen azalea with pale flowers
that have pink- or lavender-edged petals.
H 0.6m (2ft); S 0.6m (2ft).

2 *Leucothoe* 'Scarletta' **Leucothoe**
Leaves are dark red-purple in colour when young
but turn dark green in summer then bronze in
winter. H 1–2m (3–6ft); S 3m (10ft).

3 *Ceanothus* 'Concha' **California lilac**
Large evergreen shrub covered with dense, heavy
clusters of vivid, dark-blue flowers in late spring.
H 3m (10ft); S 3m (10ft).

4 *Paeonia delavayi* **Tree peony**
Tall, upright deciduous shrub form of peony with
dark-red flowers that can each grow to 10cm (4in)
in diameter.
H 2m (6ft); S 1.2m (4ft).

5 *Kerria japonica* 'Golden Guinea' **Japanese rose**
Lovely, bright spring-flowering shrub with broad,
hypericum-like golden-yellow flowers.
H 2m (6ft); S 2.5m (8ft).

6 *Photinia* x *fraseri* 'Red Robin' **Photinia**
Large evergreen shrub that can grow to the size
of a small tree. New leaves are scarlet-red, but
turn green as they age.
H 5m (15ft); S 5m (15ft).

7 *Chaenomeles* x *superba* 'Crimson and Gold'
Flowering quince
Thorny, deciduous shrub that produces bright
red flowers followed by yellow fruits.
H 1m (3ft); S 2m (6ft).

8 *Kolkwitzia amabilis* 'Pink Cloud' **Beauty bush**
Deciduous shrub that in late spring is completely
covered in masses of deep-pink flowers.
H 3m (10ft); S 4m (12ft).

SHRUBS FOR SUMMER AND AUTUMN COLOUR

1 *Potentilla fruticosa* Princess **Shrubby cinquefoil**
Low-growing, deciduous shrub with long-lasting pale-pink flowers that may fade to white in the sun.
H 60cm (2ft); S 1m (3ft).

2 *Kalmia latifolia* **Calico bush**
A dense bushy shrub bearing distinctive pink flowers from crimped buds in summer.
H 3m (10ft); S 3m (10ft).

3 *Hydrangea macrophylla* 'Maculata' **Hydrangea**
Lacecap, commonly sold as 'Blue Wave'; the blue colour of the flowers is only reliable on acid soils.
H 2m (6ft); S 2.5m (8ft).

4 *Helianthemum* 'Wisley Primrose' **Rock rose, sun rose**
Compact evergreen shrub with grey-green leaves and in summer, pale yellow flowers with darker centres.
H 20–30cm (8–12in); S 30–45cm (12–18in).

5 *Phlomis fruticosa* **Jerusalem sage**
Evergreen shrub with grey-green, sage-like leaves and clusters of hooded, golden-yellow flowers.
H 1m (3ft); S 1.5m (5ft).

6 *Euonymus europaeus* 'Red Cascade' **Spindle tree**
Deciduous shrub or small tree with an extraordinary mixture of scarlet-crimson leaves and fuchsia-pink fruits in autumn.
H 3m (10ft); S 2.5m (8ft).

7 *Fuchsia* 'Andrew Hadfield' **Fuchsia**
Small shrub with generous displays of single carmine-red flowers through summer and into autumn.
H 20–45cm (8–18in); S 20–30cm (8–12in).

8 *Leycesteria formosa* **Himalayan honeysuckle**
A thicket-forming, deciduous shrub with green stems. Drooping flowers are followed by purple berries.
H 2m (6ft); S 2m (6ft).

SHRUBS FOR WINTER COLOUR

1 *Mahonia* x *media* 'Charity' **Mahonia**
Upright and spiny-leaved evergreen, architectural, with large, spiky clusters of scented flowers.
H to 3m (10ft); S to 2m (6ft).

2 *Leucothoe fontanesiana* 'Rainbow' **Switch ivy**
Evergreen whose dark green leaves are mottled cream and pink in winter. Clusters of white flowers in late spring.
H 1.5m (5ft); S 2m (6ft).

3 *Cornus alba* 'Sibirica' **Dogwood**
A deciduous shrub with bright red leaves in autumn and even more vibrant red stems in winter.
H 3m (10ft); S 3m (10ft).

4 *Daphne bholua* **Daphne**
Upright shrub bearing clusters of fragrant pinkish-white flowers in winter, followed by fleshy fruits.
H 5–15cm (2–6in); S 1m (3ft) or more.

5 *Skimmia japonica* 'Bronze Knight' **Skimmia**
Vigorous evergreen shrub with dark, glossy pointed leaves and clusters of dark-red winter buds.
H to 6m (20ft); S to 6m (20ft).

6 *Ilex aquifolium* 'Madame Briot' **Holly**
Bushy, evergreen shrub or small tree with spiny, dark-green leaves edged with gold and bright red berries.
H 10m (30ft); S 5m (15ft).

7 *Euonymus fortunei* 'Emerald 'n' Gold'
Evergreen bittersweet
A bushy, evergreen shrub bearing white-edged, bright green leaves, tinged pink in winter.
H 60cm (24in); S 90cm (36in).

8 *Hamamelis* x *intermedia* 'Jelena' **Witch hazel**
Tall deciduous shrub with orange flowers on bare branches in midwinter. Good autumn leaf colour, too.
H 4m (12ft); S 6m (20ft).

EVERGREEN SHRUBS

1 *Viburnum tinus* 'Variegatum' **Viburnum**
Dense, compact shrub with pointed green leaves
edged with creamy yellow. Clusters of red flowers
in late winter and spring.
H 3m (10ft); S 3m (10ft).

2 *Pittosporum tobira* **Pittosporum**
Dark-green, glossy leaves are dense enough for
a line of shrubs to act as a windbreak hedge.
H 2–10m (6–30ft); S to 6m (20ft).

3 *Fatsia japonica* **Japanese aralia**
Large, architectural plant with huge, hand-shaped
glossy green leaves. Clusters of white flowers
followed by black berries.
H 1.5–4m (5–12ft); S 1.5–4m (5–12ft).

4 *Origanum* 'Kent Beauty' **Marjoram** or **oregano**
Low-growing, trailing subshrub. In summer
produces pale- and rose-pink flowers, rather
like hops.
H 10cm (4in); S 20cm (8in).

5 *Lavandula stoechas* **Lavender**
Compact, mound-forming shrub with long,
thin, grey-green leaves. Dark purple flower spikes
with distinctive "ears" or bracts.
H 60cm (2ft); S 60cm (2ft).

6 *Laurus nobilis* f. *angustifolia* **Sweet bay** or
bay laurel
Large shrub or small tree that can be clipped
into formal topiary shapes. Shiny oval leaves
can be used as a herb.
H to 12m (40ft); S to 10m (30ft).

7 *Erica carnea* 'Vivellii' **Heath**
Low-growing, spreading evergreen with purple-pink
flowers from late winter to spring. Makes a good
ground cover plant.
H 15cm (6in); S 35cm (14in).

8 *Euphorbia characias* **Milkweed** or **spurge**
An upright shrub, and a striking architectural feature
plant. Flowers in spring and early summer.
H 1.2m (4ft); S 1.2m (4ft).

SHRUBS FOR FRAGRANCE

1 *Viburnum* x *burkwoodii* 'Anne Russell' **Viburnum**
Compact deciduous shrub whose clusters of pink buds produce fragrant white flowers in mid- to late spring.
H 2m (6ft); S 1.5m (5ft).

2 *Trachelospermum jasminoides* **Star jasmine**
Evergreen, woody climber with oval leaves, and fragrant white flowers in summer.
H to 9m (29ft 6in); S to 9m (29ft 6in).

3 *Sarcococca confusa* **Christmas box** or **sweet box**
Dense, dark green evergreens that bloom in winter with intensely fragrant white flowers.
H 2m (6ft); S 1m (3ft).

4 *Choisya* x *dewitteana* 'Aztec Pearl' **Mexican orange blossom**
Compact evergreen; scented flowers in late spring, and again in late summer, especially if deadheaded.
H 2.5m (8ft); S 2.5m (8ft).

5 *Osmanthus* x *burkwoodii* **Osmanthus**
Evergreen with glossy, dark green leaves and white flowers that have a fragrance similar to jasmine.
H 3m (10ft); S 3m (10ft).

6 *pink Argyrocytisus battandieri* **pink Moroccan broom**
Tall, deciduous shrub with silver-green leaves and clusters of yellow, pineapple-scented flowers in summer.
H 5m (15ft); S 5m (15ft).

7 *Philadelphus* 'Virginal' **Mock orange**
Powerfully fragrant, pure-white double flowers appear on this dark green deciduous shrub in early or midsummer.
H 3m (10ft); S 2.5m (8ft).

8 *Daphne mezereum* **Daphne**
A deciduous shrub that produces purple-pink flowers in early spring before the leaves appear. Fragrant but toxic.
H 3m (10ft); S 2.5m (8ft).

SHRUBS FOR WALLS

1 *Escallonia* 'Langleyensis' **Escallonia**
Fast-growing evergreen or semi-evergreen
with glossy leaves and pink flowers. Makes
a good hedge.
H 2m (6ft); S 3m (10ft).

2 *Fremontodendron* 'California Glory'
Flannel flower
Upright, spreading evergreen best grown
against sunny walls. Deep yellow flowers from
late spring to autumn.
H 6m (20ft); S 4m (12ft).

3 *Jasminum nudiflorum* **Jasmine**
Sprawling, straggly shrub that positively needs
wall training and regular trimming. Tolerates
north- or east-facing walls.
H 3m (10ft); S 3m (10ft).

4 *Pyracantha* 'Watereri' **Firethorn**
Spiny evergreen with white flowers and red
berries can be fan trained or espaliered against
walls or fences.
H 2.5m (8ft); S 2.5m (8ft).

5 *Euonymus fortunei* 'Emerald Gaiety'
Evergreen bittersweet
Bushy, spreading shrub with bright green
leaves whose white edges turn pink in winter.
H 1m (3ft); S 1.5m (5ft).

6 *Cotoneaster horizontalis* **Fishbone cotoneaster**
A deciduous shrub whose branches spread sideways
in a fishbone pattern. Red leaves and bright red
berries in autumn.
H 1m (3ft); S 5m (15ft).

7 *Garrya elliptica* **Garrya**
Large, pollution-resistant evergreen with tough,
leathery leaves. Male plants produce long grey-green
catkins in mid- to late winter.
H 4m (12ft); S 4m (12ft)

8 *Forsythia suspensa* **Forsythia**
Deciduous shrub with golden yellow flowers in
early to mid-spring, before leaves appear.
H 3m (10ft); S 3m (10ft).

CLIMBERS

Climbers produce long, flexible stems which they use to twine around or hook onto their support. Others cling on with specially adapted suction pads or aerial roots. Using this ability, they will scramble over plants – or anything else – to get to the best light. Use this to your advantage by planting climbers where they will disguise any eyesores in the garden, or brighten up dreary walls and fences. In a small garden, climbers are real assets, giving height without taking up too much ground space; use their foliage and flowers to dress up trellises, walls, and other garden structures, or let them climb through trees and shrubs to prolong the interest of these natural supports.

Most clematis can be trained to climb up the trunk of a tree, although some – like this *Clematis alpina* – may need some help from twine or canes until they establish their grip.

GROWING CLIMBERS

Most climbers are easy to grow. They are inherently vigorous, as it is in their nature to scramble and spread upwards and outwards in pursuit of as much light as possible. However, woody varieties can take a few years to establish. Before they reach their required height and spread, you can fill the space with annual climbers for instant results. Sweet peas, morning glories, and climbing nasturtiums are all good choices for sunny sites, although the latter also tolerate some shade. Annual climbers are also useful for embellishing a vegetable garden, as their root systems are less invasive than those of woody or perennial plants, and they will not interfere with crop production. Sweet peas are ideal here, adding scent and encouraging pollinating insects, as well as providing a bonus crop of cut flowers.

Leafy climbers are the perfect plants to create a cool retreat. Here, ivy and *Hydrangea anomala* subsp. *petiolaris* scale the house walls, softening the brickwork. Both are self-clinging, and need no additional wires or supports to climb, but be aware that they can damage old masonry.

Annual climbers such as these black-eyed Susans (*Thunbergia alata*), and also morning glories, sweet peas, and nasturtiums, are perfect for filling in gaps while a woody climber establishes, or for a wigwam support in a border.

Wisteria dripping with scented flowers is a stunning sight in spring, but remember that it is very vigorous, needs strong support, and must be pruned regularly where space is limited (*see p.127*).

CHOOSING CLIMBERS

While some climbers are quite small, reaching up to around head-height, others can extend to the top of large trees or over your house, so check the final heights of your chosen plants when making your plans. Also look out for mature specimens in other people's gardens or parks to see how they perform.

To plant a few climbers together so that they intermingle, ensure their pruning requirements are compatible. A late-blooming large-flowered clematis, such as the dark purple 'Jackmanii Superba', makes a good match with one of the viticella types, such as the rich red 'Madame Julia Correvon' or dark purple 'Étoile Violette', as both are pruned back to about

30cm (12in) in early spring (*see p.156*). But neither makes a good match with early flowering types, such as *Clematis alpina*, which can be left or pruned after flowering, from mid- to late spring. When pruning the late-flowering variety, you might remove the wrong stems and remove the flowers from your spring performer, too. Likewise, plant climbers with similar support requirements together; understanding how they climb will help you provide the best supports (*see pp.122–123*).

HOW CLIMBERS CLIMB

- **Stem twiners** (as on honeysuckle and wisteria) wind around their supports, and can be trained along wires or up posts.
- **Stem roots** (as on ivies) are produced on certain plants to help them cling to vertical supports, such as walls and fences. They can damage old walls.
- **Leaf stalk and tendril twiners** (as on clematis and sweet peas) have thin, flexible leaf stalks or tendrils that twine on contact with wires, trellis, or other plant stems.
- **Adhesive pads** (as on Virginia creepers) adhere to walls and posts, so they do not need wires, trellis or canes for support.
- **Hookers** (as on roses and brambles) use the thorns along their stems to hoist themselves up over their support.

WHERE TO PLANT CLIMBERS

In the wild, many climbers scramble up through trees and shrubs to the light at the top. To imitate these conditions in your garden, ensure the climber has its head in the sun, and use other plants around its base to shade the roots.

Climbers will also grow up established trees with light canopies, such as old fruit trees or robinias; light filters through the branches and encourages the climber to work its way up through the canopy. Good candidates for this type of display include rambling roses, such as blowsy pink 'Albertine' and creamy 'Bobbie James', and summer-flowering, large-flowered clematis such as 'Perle d'Azur', which will start to bloom as the rose finishes.

For walls and fences supporting wall shrubs or roses, fix horizontal wires 45cm (18in) apart. For leaf or stem twiners, add vertical wires as well, or use large-mesh galvanized wire.

(clockwise from top left) **Honeysuckle, clematis, roses, and Boston ivy.**

Around conifers, plant just outside the edge of the canopy, and use canes to guide the stems up to the lowest branches; plant a little closer to the trunk of deciduous trees. Try to plant on the windward side of the tree, so that unsecured young shoots are blown towards the stem.

South-facing walls can be too hot for some climbers; choose a sun-lover, such as a passion flower, wisteria, trachelospermum, or a vine (*Vitis*). The soil will also be dry next to a sunny wall, so plant about 45cm (18in) away from it; then mulch in spring, and water in dry weather.

Plant clematis and roses where they will receive the cooler morning or early evening sun, because both dislike the intense heat of a south wall, unless it is partially shaded.

SUPPORTS FOR CLIMBERS

The method by which your plant climbs (*see opposite*) will influence the type of support it requires, and its vigour will determine how large and strong the support needs to be.

On walls, the least obtrusive support to add is galvanized wire; on the other hand, fancy trellis can be used to add decorative appeal to workaday surfaces. Plastic and metal ties must be checked and loosened regularly, particularly on woody climbers. Twine or raffia, with some "give", are better.

On pergolas and arches, fix vertical wires to run up and down opposite sides of the posts. Train and tie in your climber in a corkscrew fashion when the stems are young and flexible.

Home-made bamboo cane tripods will support annuals and most perennials. Tripods, pyramids, and obelisks must be sturdy enough to withstand the weight and vigour of the climber.

PLANTING AGAINST A WALL OR FENCE

Attach supports before you prepare the soil, or it will become compacted. With freestanding supports like arches and obelisks, dig over the area before erecting them, rather than negotiating around the uprights later. Vine eyes are invaluable for attaching wires to vertical surfaces. The wire can be pulled taut by giving the vine eye a few more turns once the wire is in place.

1 Water the plant well an hour before planting, or submerge it in a bucket of water. Allow to drain. Then dig a hole about 45cm (18in) from the fence, not hard up against it in the dry "rain shadow". Check that the hole is deep enough for the plant.

2 Use bamboo canes arranged in a fan-shape and leaning into the fence to form a temporary support on which to guide the climber's stems up to the permanent wires. The canes can be removed later.

3 Lean the plant against the canes; make sure it is still level with the compost (except for clematis, which should be about 10cm (4in) below the surface level). Backfill the hole with a trowel, then firm in with your fists.

4 Untie the plant from its original cane and remove this, then fan out the main stems and tie into the canes, using soft twine in a loose figure-of-eight. Tying just above the cane's joints should prevent the twine slipping down.

5 Fluff up the soil, then draw it up to form a circular ridge around the climber's base, creating a saucer-shaped depression. Drench the soil with water. The hollow will retain the excess water, allowing it to seep into the root area. Water the new climber regularly from then onwards, especially if planting in spring.

6 Apply a mulch around the plant, but do not allow it to touch the stems. Once the stems reach the wires, train the outermost ones horizontally along the lower wires, and the central ones up and along the upper ones – this will ensure good coverage.

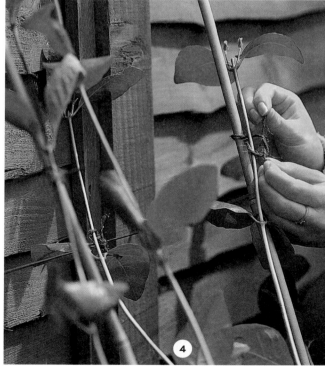

PRUNING CLIMBERS

Climbers stand on other plants' shoulders to reach the light, then hog as much of it as they can. As a result, they may grow up out of sight, leaving their trailing bare stems behind – not ideal in a small garden. An easy remedy is to cut them back severely, either in late winter, as for many late summer- or autumn-flowering clematis, or every few years, as for honeysuckle. Climbers that flower early in the year and bloom on stems made the previous year should be trimmed after flowering.

WHEN TO PRUNE CLIMBERS

Different types of climber require pruning at different times of year – often depending on when they flower and whether the flowers appear on new or old stems.

Clematis fall into three main groups: those that flower in late winter or spring; those that flower in early summer; and the later-flowering types that bloom from midsummer to autumn. For the best flowers, each should be pruned in a different way (*see below*).

Some woody climbers, including actinidia, parthenocissus, and wisteria, should be pruned in winter when dormant, like many trees and shrubs, although wisteria needs further pruning in summer for the best results (*see opposite*).

Early-flowering clematis (including *Clematis montana, C. armandii, C. cirrhosa,* and *C. macropetala*) need little pruning. After flowering, give them a light trim, cutting back over-long or unproductive stems to a healthy bud. Renovation is possible for old, straggly plants: cut back all stems almost to the ground, but do not repeat this for at least three years.

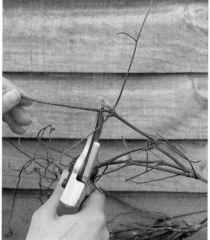

Early- to midsummer flowering clematis can be left to their own devices, unless the plant needs restricting, when a trim after flowering will help. For renovation of old tangled plants, cut back to buds near soil level in late winter. You may either lose that year's flowers, or the plant may flower later in the summer as a result.

Late-flowering clematis, which do not flower until late summer or autumn, are pruned in late winter or early spring when the buds are starting to swell. Prune them down to a pair of plump, healthy buds about 30cm (12in) from the base. Alternatively, if you are dealing with a vigorous type, it can be cut right down to the ground.

PRUNING AN ESTABLISHED WISTERIA

Wisteria is a special case, where shortening the sideshoots promotes the formation of flower buds. You will need to prune twice: once in summer, and again when the plant is dormant in winter. For large plants, you may need a sturdy ladder to reach the top.

1 In summer, after flowering, prune back the sideshoots to within five to seven leaves from the main stem, and tie in new growth.

2 In winter, shorten the new stems that have formed after you pruned in summer. Take these back to leave two buds. The short sideshoots that are left will produce lots of spring blooms.

Pyracantha can be trained up a wall or a fence like a climber. New berries are formed at the base of stems made the previous year, so take care not to prune them off.

PRUNING AND TRAINING WALL SHRUBS

Certain shrubs – though not actual climbers – can be treated almost as if they were by growing them against fences, walls and trellis. Use the same pruning methods as for free-standing plants, but tie in stems during the summer in order to guide growth over the surface to be covered. At the same time, shorten any shoots that are growing away from the wall to make sure the shrub stays neat and narrow.

Pyracantha, cotoneaster, and ceanothus make excellent wall shrubs; all flower in spring at the base of stems made the previous year. Make sure that the wall is in good condition before fixing trellis to it, as any repairs will be difficult once the plant is established. When pruning, take care not to cut back too hard into the old wood if you want more blooms and berries. Shorten all new sideshoots in midsummer, leaving two to three leaves to encourage flower formation the following year, and to make any berries more visible.

CLIMBERS FOR FLOWERS AND FOLIAGE

1 *Tropaeolum speciosum* **Flame nasturtium**
A perennial, climbing variety of nasturtium with vermilion flowers in summer followed by small blue berries.
H 3m (10ft).

2 *Solanum crispum* 'Glasnevin' **Solanum**
Fast-growing climber with small, oval green leaves and clusters of purple-blue flowers with yellow centres in summer.
H 6m (20ft).

3 *Rhodochiton atrosanguineus* **Rhodochiton**
A leaf-stalk-twining climber with heart-shaped, green leaves and dark purple and red pendant-like flowers.
H 3m (10ft).

4 *Vitis coignetiae* **Vine**
Relative of the grapevine with unpalatable fruits. Grown for its huge leaves and vibrant autumn colour.
H 15m (50ft).

5 *Eccremocarpus scaber* **Chilean glory flower**
Fast-growing climber with dramatic tubular flowers; grows well as an annual in temperate climates and may overwinter in warm areas.
H 3–5m (10–15ft).

6 *Hedera helix* 'Eva' **Ivy**
A variegated ivy with small but dense foliage heavily splashed with white. Good for lightening a gloomy wall.
H 1.2m (4ft).

7 *Actinidia kolomikta* **Actinidia**
The pink- and white-splashed leaves are the main attraction, although on young plants leaves are green.
H 5m (15ft).

8 *Passiflora caerulea* 'Constance Elliott' **Passion flower**
Fast-growing climber with long, green leaves and exotic, scented white flowers in summer and autumn.
H 10m (30ft).

CLIMBERS FOR SCENT

1 *Jasminum officinale* 'Argenteovariegatum' **Jasmine**
Late summer-flowering, hardy, twining jasmine, with variegated leaves edged in cream.
H 12m (40ft).

2 *Wisteria floribunda* 'Alba' **Wisteria**
Large, woody, deciduous climbers that can live for many years once established. White, hanging flower clusters up to 60cm (2ft) long.
H 10m (30ft).

3 *Lonicera periclymenum* 'Serotina' **Honeysuckle**
Powerfully scented flowers in mid- and late summer. Both flowers and berries are very popular with wildlife.
H 7m (22ft).

4 *Solandra maxima* **Chalice vine** or **cup of gold**
Vigorous evergreen climber with fragrant, trumpet-shaped, golden-yellow flowers. Needs a warm spot.
H 7–10m (23–30ft).

5 *Trachelospermum jasminoides* **Star jasmine**
Woody evergreen that needs a sunny, sheltered wall. Pure white, scented flowers in mid- to late summer.
H 9m (28ft).

6 *Lathyrus odoratus* **Sweet pea**
Vigorous tendril climbers available in hundreds of varieties. Flowers in shades of red, pink, mauve, blue, and white – often scented.
H 2m (6ft).

7 *Akebia quinata* **Chocolate vine**
Large, twining, woody climber with spice-scented, cocoa-coloured flowers sometimes followed by long fruits.
H 10m (30ft).

8 *Beaumontia grandiflora* **Herald's trumpet**
The shape of the scented, white flowers gives this evergreen climber its common name. It flowers from late spring to summer.
H 5–15m (15–50ft).

CLIMBERS FOR SHADY WALLS

1 *Hedera helix* 'Dragon Claw' **Ivy**
Most ivies are suitable for a cold wall. This quick-growing, curly-leaved variety is also known as 'Curly-Q'.
H 2m (6ft).

2 *Schizophragma integrifolium* **Schizophragma**
Large, heavy, deciduous climbers related to hydrangeas, with similar foliage and clusters of creamy-white flowers.
H 12m (40ft).

3 *Clematis* 'Hagley Hybrid' **Clematis**
Vigorous, compact, summer-flowering clematis, with large, single flowers with pinkish-mauve sepals.
H 2m (6ft).

4 *Hedera hibernica* **Ivy**
A fast-growing ivy with triangular, dark-green leaves that will grow almost anywhere.
H 10m (30ft).

5 *Hydrangea anomala* subsp. *petiolaris*
Climbing hydrangea
A wall-climbing variety with large, white flowerheads and leaves that turn from green to gold in autumn.
H 15m (50ft).

6 *Lonicera* x *americana* **Honeysuckle**
Deciduous honeysuckle that grows well in shade. Extremely fragrant flowers followed by small, red berries.
H 7m (22ft).

7 *Hedera pastuchovii* subsp. *cypria* **Ivy**
A wall-climbing ivy originally from Cyprus with triangular green leaves marked with grey-green veins.
H 3m (10ft).

8 *Lathyrus latifolius* **Everlasting pea**
Perennial climber with slender, blue-green leaves and strings of small, pink-purple flowers.
H 2m (6ft).

FAST-GROWING CLIMBERS

1 *Allamanda cathartica* **Golden trumpet**
Vigorous evergreen grown largely for its bright yellow flowers from summer to autumn.
H 8–15m (25–50ft).

2 *Ipomoea coccinea* **Morning glory**
A quick-growing, twining climber grown as an annual. Bright red and yellow flowers in summer.
H 3m (10ft).

3 *Passiflora caerulea* **Blue passion flower**
Vigorous climber that flowers from summer into autumn, then produces edible but uninteresting, orange-yellow fruits.
H to 10m (30ft).

4 *Fallopia baldschuanica* **Fallopia**
Also known as the 'mile-a-minute plant', its name says it all. Useful for quickly covering ugly walls – or entire buildings.
H 12m (40ft).

5 *Parthenocissus tricuspidata* **Boston ivy**
Vigorous form of Virginia creeper whose leaves turn wonderful fiery shades of orange, red and purple in autumn.
H 20m (70ft).

6 *Cobaea scandens* f. *alba* **Cup and saucer vine**
A wall-climbing variety with large, white flowerheads and leaves that turn from green to gold in autumn.
H 15m (50ft).

7 *Humulus lupulus* 'Aureus' **Golden hop**
A perennial climber with beautiful golden-yellow foliage and scented flowers in summer.
H 6m (20ft).

8 *Parthenocissus henryana* **Chinese Virginia creeper**
Long, slender leaves are green with clearly marked grey or even pinkish veins; they turn bright-red before falling in autumn.
H 10m (30ft).

EARLY- AND MIDSUMMER-FLOWERING CLEMATIS

1 *Clematis* **'Vyvyan Pennell'**
Unusual, double flowers with a lilac-coloured, velvety texture in early summer.
H to 2.5–3m (8–10ft).

2 *Clematis* **'White Swan'**
Compact variety with beautiful, semi-double, white flowers. Blooms in late spring.
H to 2m (6ft).

3 *Clematis armandii* **Armand clematis**
Strong-growing evergreen with tapering, glossy, green leaves and fragrant, white flowers in spring.
H 3–5m (10–15ft).

4 *Clematis* **'Blue Ravine'**
Large, open, violet-blue flowers. Blooms in late spring and again in late summer.
H to 3m (10ft).

5 *Clematis macropetala* **Downy clematis**
Deciduous, hardy clematis whose blue or violet, semi-double flowers appear from spring to early summer.
H to 2–3m (6–10ft).

6 *Clematis* **'Niobe'**
Large, single, deep red flowers up to 15cm (6in) in diameter throughout summer.
H to 3m (10ft).

7 *Clematis* **'Fireworks'**
Very large, 20cm (8in) purple flowers with dark magenta bars down the centre of each petal.
H 3m (10ft).

8 *Clematis montana* var. *grandiflora*
Golden clematis
Very vigorous with dark-green leaves and large white flowers in late spring and early summer.
H 10m (30ft).

LATE-FLOWERING CLEMATIS

1 *Clematis* **'Bill MacKenzie'**
Perhaps the best-known yellow clematis. Open, bell-shaped flowers followed by fluffy seedheads.
H to 7m (22ft).

2 *Clematis* **'Polish Spirit'**
Single, velvety, rich purple-blue flowers bred by Brother Stefan Franczak, a Jesuit monk from a Warsaw monastery.
H 4m (12ft).

3 *Clematis* **'Perle d'Azur'**
Small, azure-blue flowers from midsummer to autumn.
H 3m (10ft).

4 *Clematis rehderiana* **Nodding virgin's bower**
Shade-tolerant clematis with unusual tubular, scented, yellow flowers from midsummer to late autumn.
H 7m (22ft).

5 *Clematis* **'Huldine'**
Cup-shaped, white and mauve flowers appear after the height of summer and last into autumn.
H 3–5m (10–15ft).

6 *Clematis* **'Jackmanii'**
Late-flowering variety with velvety, deep purple flowers. A traditional cottage-garden classic.
H 3m (10ft).

7 *Clematis* **'Purpurea Plena Elegans'**
An old variety with rosy purple, double flowers that last for several weeks.
H 3m (10ft).

8 *Clematis tangutica* **Golden clematis**
Unusual among clematis in having yellow flowers. Late-blooming, with conspicuous fluffy grey seedheads.
H 6m (20ft).

ROSES

Guaranteed to bring a note of romance into your garden, the rose has, for many centuries, been one of the most sought-after garden flowers. With thousands of different types to choose from, think first about how you want to use yours. There are dwarf roses for patios and pots, large shrubs to fill gaps at the back of the border, and giant ramblers that will scramble over buildings and pergolas, or scale trees. Consider flowering times, too; you can either opt for roses that flower once a year and usually bear fruits, or others that bloom throughout summer.

A large shrub rose such as 'Buff Beauty', with its wonderful apricot-coloured flowers, can fill a warm, sunny corner of the garden with fragrance throughout the summer.

GROWING ROSES

Traditionally, roses were grown in their own gardens, but few people today have the space to plant formal rose beds. Mixing roses with other plants prolongs the season of interest, disguises the bare stems left after pruning, and attracts insects that devour aphids. Blackspot or mildew are also less prevalent in a mixed environment. While a little blackspot or mildew will not kill the plant, they are unsightly when they affect a whole bed. Try to hide the diseased foliage using other plants rather than apply pesticides that harm the environment and wildlife.

TYPES OF ROSE

Roses come in many shapes and forms; because many are sold when dormant and pruned back, it is hard to imagine them in bloom. Get to know the different types and their pruning needs (see *pp.140–143*) before buying. As well as the main kinds described here, the flower shape may also appear on the label – single, with the central boss of stamens visible; semi-double, with an extra ruff of petals; or double, with a rounded mass of over 30 petals.

• **Shrub** roses vary in character; a few flower only once, such as some old garden roses and wilder species, while others put on a longer show. Modern shrubs like 'Sally Holmes' (*pictured*) flower repeatedly.

• **Climbers** tend to have stiff stems with either single flowers or ones held in clusters. 'Climbing Iceberg' (*pictured*) has showy double white flowers from summer through to autumn.

• **Ramblers** are vigorous climbers, with long, flexible stems and flowers that are borne in one burst in summer. 'Sander's White Rambler' (*pictured*) is rampant, with double, scented white flowers.

• **Floribunda** bush roses, such as 'The Queen Elizabeth' (*pictured*) bear sprays of flowers. Like the hybrid teas, they flower best and most continuously if you cut off the dead flowers.

• **Hybrid tea** bush roses will produce large flowers if pruned hard. 'Precious Platinum' (*pictured*) is vigorous, and bears fully double scarlet-crimson flowers.

(opposite page, clockwise from top left)
Shrub, floribunda, rambler, hybrid tea
(right) **climber**

PLANTING A ROSE

Before planting the rose, prune off any dead, diseased, or crossing stems and damaged roots. Prepare the site by digging a hole deep enough to hold the plant, and forking some well-rotted organic matter into the bottom.

1 Lay a cane across the hole to make sure the planting depth is correct. The graft union (the bulge at the base of the stem) should be 2.5cm (1in) below the soil level – this will help to prevent suckers growing.

2 Refill the hole with soil, treading the soil gently with your foot to remove any air pockets. Rake the soil, then water it thoroughly.

SHAPE AND FOLIAGE

Because many roses can be quite ungainly when not in flower, look for those with a good overall shape. Species roses, such as *Rosa xanthina* f. *hugonis*, are graceful, outwardly arching plants. 'Geranium' makes a handsome shrub, with its brick red single flowers, followed by bottle-shaped rosehips in late summer. Other shapely shrubs include 'Marguerite Hilling', with large, bright pink single flowers throughout the summer, and 'Nevada', which produces creamy flowers.

Some roses also have interesting foliage, such as *Rosa glauca*, with bluey-green leaves offset by dark red stems. A few roses have scented foliage, which can become quite powerful on warm summer evenings when the atmosphere is slightly humid. Try *Rosa primula*, the leaves of which smell of incense, or *Rosa rubiginosa* with its apple-scented foliage.

CHOOSING AND BUYING ROSES

With such a huge range of roses on offer, making a choice can be daunting. Consider how much time you want to commit to looking after your plants. As a rule, hybrid teas and floribunda roses, such as

Choose easy-care shrub roses for a low-maintenance garden. 'Alba Maxima', for example, needs little pruning, tolerates some shade, and bears masses of creamy-white blooms.

(left) **Deadhead hybrid teas** by cutting back the spent flowers to a healthy sideshoot or strong outward-facing bud.

(far left) **Deadhead floribundas** by cutting back whole trusses of flowers to a healthy sideshoot, leaf, or bud to promote new flowering stems.

Freedom and The Times Rose respectively, tend to be the most demanding, requiring careful pruning and being intolerant of crowding by other plants – this is why they were traditionally grown on in separate rose beds.

If you have less time to spare, choose slightly tougher shrub roses. These include the old garden roses, such as albas and hybrid musks; English roses, including Redouté and Cordelia; rugosa roses, which are among the toughest and most undemanding roses available; all the species roses, like *Rosa pimpinellifolia*, *R. canina*, and *R. glauca*, as well as a large range of modern shrub and groundcover roses such as 'Marguerite Hilling', Bonica, and the Flower Carpet roses in their various colours.

A problem in small gardens is the lack of sunlight, as roses are sun-lovers. It is especially important to plant roses that flower more than once in the season in a sunny, south-facing position. If your garden receives sun for only part of the day, opt for once-only flowering types. The lovely alba roses come into this group and include beauties such as the creamy-white 'Alba Maxima' and double-flowered, pale pink 'Great Maiden's Blush', both of which have the typically healthy, pale grey-green foliage that gives this group of old roses their distinctive character.

Roses are often bought by mail order, as specialist rose nurseries offer an extensive choice. The downside is that you cannot select the actual plant you want, although most of these firms have an excellent reputation and should respond immediately if you feel a rose has arrived in poor condition.

ROSEHIPS

Although some roses only flower once, in summer their blooms are followed by colourful rosehips. These add sparkle to autumn and winter gardens, as well as providing a valuable food source for birds and other wildlife. All roses produce hips, although albas, rugosas, *Rosa moyesii*, and the dog rose, *R. canina*, provide the most dramatic displays.

Hips vary in shape as well as vigour: many rugosa roses, such as *Rosa rugosa* 'Alba', 'Nyveldt's White', and 'Fru Dagmar Hastrup', bear chubby, round red hips, while others, such as the vigorous rambler *Rosa filipes* 'Kiftsgate', produce sprays of little red buttons. To preserve hips, don't deadhead the flowers, and give your roses plenty of water throughout the growing season to swell the fruit.

PRUNING BUSH ROSES

Bush roses include hybrid teas and floribundas, both of which need to be cut back hard each year to encourage new growth. But do not prune floribundas quite as ruthlessly as hybrid teas, as this will reduce the number of flowers.

1 To prune hybrid tea roses remove any shoots that are dead, diseased, weak, or crossing from the centre of the bush to leave an open symmetrical framework from which new shoots can arise.

2 Reduce the main stems to 20–30cm (8–12in) above the ground, cutting back to a healthy, outward-facing bud – hybrid tea roses respond well to drastic pruning. After a few weeks, new shoots will appear, eventually producing an abundance of blooms.

To prune floribunda roses, cut the main shoots back to about a third of their length, and trim any remaining healthy sideshoots by about one- to two-thirds.

PRUNING OLD-FASHIONED ROSES

Old-fashioned roses, such as albas, mosses, and damasks, are pruned lightly in summer after the first flush of flowers (*see right*). To renovate old specimens, cut back all but the strongest young stems to the ground, and prune those that remain, taking them back by one third of their length. Renovate after flowering or in the spring – if you choose spring, you will lose the forthcoming season's flowers.

Gallica roses are pruned in a similar way to other old-fashioned roses, but cut the sideshoots back to the main stem, or to a shoot close to the main stem.

1 Remove any wood that is dead and diseased, then cut back the main stems and sideshoots by one third to a healthy, outward-facing bud.

2 When autumn arrives, cut back any extra-long whippy stems that could be vulnerable to damage by strong winter winds.

PRUNING CLIMBERS AND RAMBLERS

Climbers and ramblers are best pruned in late summer, when you can see the different kind of shoots more easily, although they will respond well to pruning in winter, too.

Each year, ramblers make lots of new growth that flowers the following year. To prevent your rose becoming a tangled mess, cut out the oldest stems near to the ground in early autumn, and retain young stems, tying them in to replace the old ones. On the remaining older wood, shorten the sideshoots to two to four healthy buds from the main stem. The young stems will flower the following year, while more new shoots will grow from the base of the plant.

(top) **For rambling roses,** prune the sideshoots back to 2–4 healthy buds from the main stem.

(bottom) **Tie in the new growth** to wires or other supports.

PRUNING CLIMBING ROSES

After planting a climber, do not prune back the long stems. Instead tie them to canes and tie the canes in a fan shape to horizontal wires fixed to your support. Train young flexible sideshoots horizontally along the wires as they grow. When established, you can prune each year to keep the plant tidy, and to encourage plenty of flowering shoots.

1 Retain old wood, except for any diseased, dead, or weak growth, which should be cut down to the ground or to a healthy bud.

2 Only prune the sideshoots, reducing them to about two-thirds of their original length, and cutting to a bud facing in the right direction.

3 Tie in the newly-pruned stems horizontally to their support. This encourages more flowering sideshoots to form along the stems.

4 Prune back any over-long stems that are protruding beyond the support, and sideshoots growing away from it.

ROSES FOR PERGOLAS AND ARCHES

1 *Rosa* **'Albertine'**
A popular rambler with arching, branching stems carrying clusters of coppery-pink, scented flowers. H 5m (15ft); S 2.5m (8ft).

2 *Rosa* **'Phyllis Bide'**
A vigorous climber for training up walls or pillars. Small, double flowers are buff-pink flushed with yellow. H 2.5m (8ft); S 1.5m (5ft).

3 *Rosa* **'New Dawn'**
Easy to grow; will tolerate a north-facing site. Fragrant, pale, blush-pink flowers in summer. H 5m (15ft); S 5m (15ft).

4 *Rosa* **Handel**
Modern climber, quite compact, and lovely for an obelisk or arch. Makes good new growth from the base each year. H 3m (10ft); S 2.2m (7ft).

5 *Rosa* **Laura Ford**
Upright climber with dense clusters of semi-double, bright yellow flowers touched with pink. H 2.2m (7ft); S 1.2m (4ft).

6 *Rosa* **'Compassion'**
Strongly scented, bushy climbing rose with dark-green leaves and double blooms in salmon-pink tinted with apricot. H 3m (10ft); S 2.5m (8ft).

7 *Rosa* **'Climbing Iceberg'**
A climbing version of the popular floribunda bush rose. Plenty of sweet-scented flowers, and not too thorny. H 4m (12ft); S 3m (10ft).

8 *Rosa* **Summer Wine**
Strong, upright climber with large, semi-glossy leaves and coral-pink flowers that bloom in summer. H 3m (10ft); S 2.2m (7ft).

ROSES FOR MIXED BEDS

1 *Rosa* **Baby Love**
Dwarf modern bush rose with small, bright-yellow flowers. Excellent for the front of a border or as a low hedge.
H 80cm (32in); S 60cm (24in).

2 *Rosa* **'Ballerina'**
Compact shrub rose with multiple clusters of single, pale-pink flowers with white centres.
H 1–1.2m (3–4ft); S 1.2m (4ft).

3 *Rosa* **Bonica**
A low shrub rose, excellent for ground-cover; long flowering season followed by bright red hips.
H 1m (3ft); S 1.2m (4ft).

4 *Rosa* **'Roseraie de l'Haÿ'**
Rugosa rose with wrinkled leaves; a large, dense shrub. Flowers almost continuously until late autumn.
H to 2m (6ft); S 2m (6ft).

5 *Rosa* **'Madame Hardy'**
Vigorous Damask rose with dark green leaves and fully double, scented, white blooms, each with a distinctive green eye.
H 1.5m (5ft); S 1.2m (4ft).

6 *Rosa* **'Constance Spry'**
Vigorous, sprawling shrub rose that needs hard pruning or training as a climber. Large, myrrh-scented, pink flowers.
H 2m (6ft); S 1.5m (5ft).

7 *Rosa* **'Complicata'**
Large, old-fashioned Gallica shrub rose with cascades of open, pink flowers. Good in wilder, more informal gardens.
H to 2.2m (7ft); S to 2.5m (8ft).

8 *Rosa* **'Charles de Mills'**
Suitable for hedging at the back of a border. An upright Gallica rose with large, fully double, magenta-pink flowers.
H 1.2m (4ft); S 1m (3ft).

ROSES FOR SCENT

1 *Rosa* **'Arthur Bell'**
Bush rose with bright green, glossy leaves and clusters of fragrant, double, yellow flowers.
H 1m (3ft); S 60cm (24in).

2 *Rosa* **Breath of Life**
A climber with large, scented, fully double flowers coloured a pinkish-apricot.
H 3m (9ft); S 2.2cm (7ft).

3 *Rosa* **Gertrude Jekyll**
An English shrub rose, in the old garden style but repeat-flowering. Deadhead for a continuous show of flowers.
H 1.5m (5ft); S 1m (3ft).

4 *Rosa* **Valencia**
A hybrid tea, vigorous and usually healthy, with long stout stems that make it excellent for cutting. Very strong, sweet scent.
H 75cm (30in); S 65cm (26in).

5 *Rosa* **'Félicité Parmentier'**
Old garden rose in the alba group, with a single flush of beautiful, fragrant, pale-pink flowers.
H 1.3m (4ft 6in); S 1.2m (4ft).

6 *Rosa* **Baronne Edmond de Rothschild**
A hybrid tea bush rose that produces large, scented, fully double, ruby-red flowers with petals that have pale pink undersides.
H 1m (3ft); S 75cm (30in).

7 *Rosa* **'Penelope'**
Bushy shrub rose with large clusters of scented, pale creamy-pink blooms from summer to autumn.
H 1m (3ft); S 1m (3ft).

8 *Rosa* **Fragrant Cloud**
Large, vigorous hybrid tea. The scented flowers, which fade to purple with age, are often borne in clusters.
H 75cm (30in); S 60cm (24in).

ROSES FOR PATIOS

1 *Rosa* **'Nathalie Nypels'**
A modern bush rose bearing clusters of sweet-scented, rose-pink flowers.
H 75cm (30in); S 60cm (24in).

2 *Rosa* **Laura Ashley**
A miniature climber with mauve and pink flowers. Sometimes used as ground cover, but also ideal for a hanging basket.
H 60cm (24in); S 1.2m (4ft).

3 *Rosa* **Anna Ford**
A dwarf bush rose perfect for containers. Clusters of semi-double, rich orange-red flowers with yellow centres.
H 45cm (18in); S 45cm (18in).

4 *Rosa* **Cider Cup**
Miniature floribunda rose with clusters of warm orange-apricot-pink flowers, dark leaves, and a fruity scent.
H 75cm (30in); S 60cm (24in).

5 *Rosa* **Ferdy**
A cluster-flowered shrub rose. Masses of small, double, coral-pink flowers throughout summer.
H 80cm (32in); S 1.2m (4ft).

6 *Rosa* **Baby Masquerade**
Dense, miniature bush rose with small, rosette-shaped flowers that start off yellow then turn pink and crimson.
H 40cm (16in); S 40cm (16in).

7 *Rosa* **Little Bo-peep**
An excellent miniature rose for containers, dense and bushy, and covered with flowers all season.
H 80cm (32in); S 1m (3ft).

8 *Rosa* **Warm Welcome**
A miniature climber, good on a wigwam or obelisk-style support in a tub, flowering from top to bottom.
H to 2.5m (8ft); to 1.5m (5ft).

GRASSES, BAMBOOS & FERNS

Planting grasses, bamboos, and ferns is a way of introducing striking and sometimes unexpected architectural elements into a garden. What these plants may lack in terms of brightly coloured flowers, they make up for in shape and texture. Grown either singly or in groups, they create form and structure. They add movement, too, as their tall stems, canes, fronds, and flowerheads sway and rustle in the wind. And they are certainly not monochrome; they display a huge range of colour, from an infinite variety of subtly different greens to blue, grey, silver, bronze, and gold. Don't overlook them.

Bamboo canes are sometimes reason enough on their own for choosing and growing the plant. The canes of *Phyllostachys aureosulcata* f. *spectabilis* are a spectacular golden-yellow.

GROWING GRASSES

For many centuries, grasses have been a key component in European gardens, although until relatively recently these were limited to lawns or meadows. Since the latter part of the 20th century, ornamental grasses have been used more and more in our gardens. Even though they do not have the pretty, colourful flowers of most garden plants, they bring structure, height, texture, autumn colour, winter silhouettes, and – most important of all – movement. Some, such as *Stipa gigantea* and molinias, stand well into autumn with their stems glowing golden orange, and grasses like miscanthus and calamagrostis remain attractive throughout the winter months and well into the spring.

Sound and movement add an extra dimension to a bed or border when grasses rustle in the wind. Intersperse them among traditional flowering plants for contrasts of colour and shape.

The winter garden will look much more interesting if it has some grasses, because they don't die down in autumn like other herbaceous plants. The birds can also feast on the seedheads.

CHOOSING GRASSES

Grasses usually fall into one of two groups: clumpers or spreaders. Some naturally form perfectly behaved clumps, whereas others can take over your garden with their creeping rootstocks. So choose carefully. There is now a large selection of grasses available that do not self-seed or spread wildly. If in doubt, check with your supplier, or take a close look at how the plant is growing. If it makes a neat, dense clump or tussock, it should not run; if you see shoots coming up all over the pot without forming a tight clump, beware. On light sandy soils in particular, grasses, such as variegated gardener's garters (*Phalaris arundinacea*), can become invasive and need to be regularly weeded out.

WHERE TO PLANT GRASSES

You can use grasses as solitary feature plants, or scattered through other plants, or in larger drifts or swathes. How you do it depends on the type of grass you choose and the effect you wish to achieve. Try large drifts of *Miscanthus sinensis* 'Kleine Silberspinne' – although not the tallest grass, it is still impressive. *Calamagrostis* x *acutiflora* 'Karl Foerster' also looks good in a mass. You could use it to create a hedge-like screen, or add depth to a bed or border by scattering it through a planting of smaller flowering perennials.

Since many grasses stand through autumn, you can create a beautiful effect by siting them carefully. The flower spikes and orange stems of *Molinia caerulea* subsp. *arundinacea* 'Windspiel' or 'Transparent' glow if they catch the late afternoon autumn sun. As its name suggests, 'Transparent', like other molinias, is almost translucent. If you position it at the front of a planting, it creates a see-through screen, and the feathery plumes brush your face as you walk past.

PLANTING GRASSES

The ideal time to plant grasses is during spring or autumn when the ground is warm; winter is sometimes an option, but only if the soil isn't rock hard or waterlogged. Water a plant well before planting by placing the pot in a bucket of water. Prepare a hole four times larger than the pot; sit the pot in the hole to check that the soil level will be at the same height once planting is complete.

1 Carefully slide the plant out of its pot. Tease out some of the roots to help the plant establish successfully in the soil.

2 Top up the planting hole with soil, and use your foot to firm it. Then water the plant thoroughly.

NATURALISTIC GRASS PLANTING

Designers in Germany and the United States have developed a way of planting perennials and grasses that mimics the plants' natural growth habits. Plants are chosen from a natural habitat with conditions similar to those in the garden. For example, if your garden is partly shaded, you should choose plants that originate from woodland margins, which will thrive in light shade with a few hours of sun each day.

Plants are placed how they would occur naturally. For example, grasses with creeping or spreading roots are planted in drifts and groups. Those that scatter themselves by means of seed are dotted about. Any bare soil is covered with ground-cover plants or a mulch to suppress weeds. The result is a community of plants that are allowed to grow and evolve naturally. Looking after such a scheme involves less maintenance than traditional gardening, since plants are not supported or deadheaded, but simply tidied up when they are past their best.

KEEPING GRASSES LOOKING GOOD

Ornamental grasses need little attention to keep them looking good. Most evergreen grasses, including sedges and rushes, only need a trim to remove damaged leaves and old flower stems. The best time to do this is in late winter, before fresh spring growth emerges. Grasses that turn straw-coloured in autumn should be cut down to the ground in late winter, after you have enjoyed their frosted foliage. New growth will soon emerge.

Most grasses will grow in poor soils and thrive on neglect – give them too much water, food, and fuss and they'll die. Large grasses, such as pampas, are an exception, and need an occasional feed with a general-purpose fertilizer to keep them vigorous. Grasses seldom need watering, except after planting and while they establish, but if their leaves roll up, it's a sign they need a drink.

Clear debris, such as old stems, using a spring-tined, or grass, rake. Work from the centre of the clump and comb out any rubbish.

Gravel is an attractive and effective mulch for grasses; it will conserve moisture and suppress weeds.

Cut back old stems close to the ground, using secateurs or shears. If the grass has sharp edges, wear eye protection and gloves.

Use loppers on thicker grasses, such as miscanthus, where the stems are more like canes. You may even need to use a saw.

GRASSES FOR LEAF COLOUR

1 *Festuca glauca* **Blue fescue**
Dense, tufted evergreen grown for steely blue leaves.
Flowers in early to midsummer.
H to 30cm (12in); S 25cm (10in).

2 *Holcus mollis* 'Albovariegatus' **Striped Yorkshire fog**
A creeping, perennial grass with flat, soft leaves
in blue-green with creamy white edges. Forms
a carpet-like ground cover.
H 20cm (18in); S 45cm (18in).

3 *Miscanthus sinensis* 'Variegatus'
Variegated miscanthus
Tall grass with narrow, variegated leaves in
pale-green and creamy white stripes.
H to 1.8m (6ft); S 25cm (10in).

4 *Elymus magellanicus* **Magellan rye grass**
Dense, tufted perennial grass. Sometimes produces
flat flower spikes in summer and autumn.
H to 1m (3.2ft); S to 0.5 m (1.6ft).

5 *Molinia caerulea* subsp. *caerulea* 'Variegata'
Striped purple moor grass
Small, deciduous grass whose leaves are striped
cream and turn clear yellow in autumn. Purple
flower spikes.
H 45–60cm (18–24in); S 40cm (16in).

6 *Hakonechloa macra* 'Aureola' **Hakone grass**
Dense mounds of brightly coloured, pale-green
leaves flushed with red in autumn. Good for growing
in containers.
H 35cm (14in); S 45cm (18in).

7 *Helictotrichon sempervirens* **Blue oat grass**
Forms large, dense, rounded tussocks of evergreen,
tightly rolled, silvery blue-green leaves.
H to 1.4m (4ft); S 60cm (24in).

8 *Miscanthus sinensis* 'Zebrinus' **Zebra grass**
Horizontal white or yellow bands appear on the
leaves in summer. Maroon or copper-coloured
flower plumes.
H to 1.2m (4ft); S 1.2m (4ft).

GRASSES FOR FLOWERS

1 *Pennisetum setaceum* 'Purpureum' **Fountain grass**
Tropical, deciduous grass often grown as an annual.
Dark purple leaves and crimson flowers.
H 60cm (24in); S 60cm (24in).

2 *Stipa gigantea* **Giant feather grass** or
golden oats
Upright stems carry spectacular, oat-like flowers
that ripen from purple-green to gold.
H to 2.5m (8ft); S 1.2m (4ft).

3 *Miscanthus sinensis* 'Grosse Fontäne' **Miscanthus**
Its name translates as "large fountain". Tall flower
spikes in midsummer; the long, arching leaves turn
golden-brown in autumn.
H 1.7m (5ft); S 1.2m (4ft).

4 *Pennisetum alopecuroides* 'Hameln'
Fountain grass
Small, evergreen grass with distinctive, bristly
flowerheads often compared to foxtails
or bottlebrushes.
H 50cm (20in); S 50cm (20in).

5 *Cortaderia selloana* 'Silver Comet' **Pampas grass**
Slightly kitsch perhaps, but dramatic nevertheless.
This variety has large white plumes and white
striped leaves.
H 2.5–3m (8–10ft); S 1.5m (5ft).

6 *Deschampsia cespitosa* 'Goldtau' **Tufted
hair grass**
Compact variety with a cloud-like mass of tiny
flowers that change colour from silvery-red to
golden yellow in late summer.
H to 75cm (30in); S to 75cm (30in).

7 *Panicum virgatum* 'Dallas Blues' **Blue switch grass**
A perennial prairie grass originally from Texas.
Blue-green foliage fades to golden-yellow in autumn.
Large heads of delicate flowers.
H 2.5–3m (8–10ft); S 1.5m (5ft).

8 *Briza maxima* **Greater quaking grass**
Despite its name, quite a compact grass.
Grown for its unusual, hop-like flowerheads.
H 45–60cm (18–24in); S 25cm (10in).

GROWING BAMBOOS

As relatives of grasses, bamboos are not strictly woody plants, but
from a gardening point of view, we tend to treat them as such because
their tough woody canes live for many years. The absence of flowers, fruit, or
other seasonal features is made up for by their graceful, swaying stems, and
fine leaves that rustle in the breeze. The canes of some are unusually coloured,
ranging from grey-green and yellow to inky black. As a bonus, after a few
years you can harvest your own bamboo canes to use as plant supports.

WHERE TO PLANT BAMBOOS

Bamboos range in height from relatively small
ones at 10–15cm (4–6in) tall, to giants that will tower
above a house. They also vary considerably in terms
of hardiness. The majority prefer moist soil, although
Pseudosasa japonica tolerates shallow soil over
chalk, while most fargesia, sasa, and phyllostachys
tolerate shade.

Some, such as pleioblastus, tend to spread
rampantly, but others, such as phyllostachys,
remain in a manageable clump; ask for advice when
purchasing, and if choosing an invasive type, make
sure you have sufficient space, or use a root barrier.
Bamboos often spread less in moist situations than
in dry conditions.

The majority of bamboos die after flowering, but
don't let this put you off growing them, as it may not
happen in your lifetime. You can also avoid this by
buying young plants, such as offspring of *Fargesia
murielae*, which last flowered in the 1990s. Bamboo
plants of the same species are often clones, so they
will have identical lifespans throughout the world. It
is a good idea to investigate the expected life cycles
of the species before buying one.

An elegant backdrop where space is tight, *fargesia*
form an erect clump rather than spreading over and
shading out any shrubs and plants in front of them.

Bamboo screens lend a contemporary look to modern garden designs. Here, their crossed stems and feathery foliage are made more dramatic by the red background and uplighting.

A see-through bamboo screen creates a delicate divide, segregating the garden into two distinct areas. The benefit over a wall or a fence is that the bamboo stems don't block out as much light as they would planted thickly.

BAMBOO SCREENS AND BACKDROPS

Bamboos are a versatile group of garden plants. Architectural types, such as the elegant *Fargesia murielae* and *Fargesia nitida*, or the fuzzier-looking *Chusquea culeou*, look spectacular both as a main focal point in a bed, or standing alone in a hard landscape setting, such as a gravel or courtyard garden. As well as this, bamboos are particularly suited to modern designs. Try planting one species *en masse* in geometric-shaped beds in order to create a contemporary look, or use them in a row to create a dramatic screen, either as a boundary or to divide up the garden into separate rooms. For a screen or feature, consider bamboos with interesting stems, such as the black-caned *Phyllostachys nigra* or one of the yellow-stemmed types, such as *Phyllostachys bambusoides* 'Castilloni'.

BAMBOO EFFECTS

Although not strictly woody plants – bamboos are actually perennials – their permanent woody canes mean that they can be used like shrubs in the garden. The tall types, such as *Phyllostachys* or *Fargesia*, create elegant screens when planted in a row, and shorter bamboos, including sasa and *Pleioblastus variegatus*, make good groundcover. Most bamboos, except very tall varieties, work well in containers, although many enjoy damp soil conditions and thus need frequent watering when grown in pots. Alternatively, use bamboos in geometric beds in contemporary gardens, or to define and enhance Japanese or Oriental garden designs.

BAMBOOS IN CONTAINERS

One way to prevent a bamboo from spreading is simply to keep it potted up. Always use pots big enough for the roots, and never let the plants dry out. For smaller bamboos, pots should be at least 40cm (16in) in diameter; they need to be 90cm (36in) or more across for larger plants. Containers in contemporary designs and materials work really well with bamboos.

One of the best small bamboos for containers is the light, ferny *Pleioblastus pygmaeus*, or, for something taller, try *Arundinaria gigantea* subsp. *tecta*, with its yellowish stems, which reaches about 1.8m (6ft) in height.

Feed the plants throughout the growing season (a slow-release fertilizer is ideal) and make sure they do not get too crowded by cutting out about one in three of the canes each year.

Growing bamboo in pots means that you can contain it and stop it from spreading out of control.

A simple barrier sunk into the soil around the root area will prevent the bamboo encroaching on other plants' space.

CONTROLLING A BAMBOO'S SPREAD

Some bamboos are particularly invasive, and they need to be kept in check to prevent them from taking over the rest of the garden.

There are a number of ways to control the spread of bamboos. Confining your bamboo plants to a container (*see left*) is probably the simplest method, as the roots cannot spread beyond the container walls.

If you prefer your plants to stay in the soil, construct a simple barrier around them to restrain the roots. The barrier can be made of overlapping slates, steel, or non-perishable plastic, and should be at least 40cm (16in) deep. Bury it vertically in the ground, with about 8–10cm (3–4in) protruding above the surface, so that it encircles the bamboo. If possible, seal any joins in the barrier to prevent the roots escaping.

Some gardeners prefer to dig around the clump annually using a spade, and slice off and dig out roots going beyond their allotted space (*see opposite*).

CUTTING BACK BAMBOOS

Running bamboos spread by sending out rhizomes (or underground shoots) into the surrounding soil. The way to control them is to dig a trench about 30cm (12in) wide and deep all around the clump in order to reveal the rhizomes. Then sever them close to the rootball and dig them out of the ground. When they are young and still fairly soft, you should be able to slice through them easily with a sharp spade. As they age, they harden and you may have to use secateurs or loppers.

1 Use a spade blade to slice through any rhizomes that have grown into the trench area. If they are particularly tough, it may help to sharpen the blade before you do so using a sharpening stone. Alternatively, cut them away using secateurs.

2 Remove the pieces of rhizome from the soil by tugging them out by hand. If they go particularly deep into the ground, use a fork to uncover and remove them. Either dispose of them, or pot them up to create new bamboo plants.

3 Refill the trench with the soil you removed earlier. If any further shoots start to spread out beyond the rootball area, cut them off immediately. This process will probably need to be repeated next season to keep the plant in check.

ORNAMENTAL BAMBOOS

1 *Phyllostachys aureosulcata* f. *aureocaulis*
Golden bamboo or **fishpole bamboo**
Medium-sized to large bamboo that can spread vigorously in warm climates. Upright, bright golden-yellow canes.
H 2–10m (6–30ft); S indefinite.

2 *Sasa palmata* f. *nebulosa* **Sasa palmata**
Vigorous, spreading bamboo with large, 35–40cm (14–16in) long, glossy, bright green leaves.
H to 2m (6ft); S indefinite.

3 *Phyllostachys vivax* f. *aureocaulis* **Golden Chinese timber bamboo**
A stout-stemmed, fast-growing bamboo with butter-yellow canes, sometimes green-striped. Can grow tall.
H to 10m (30ft); S indefinite.

4 *Fargesia nitida* **Fountain bamboo**
Slender, purple-green canes topped by narrow, mid-green leaves. Compact and slow growing.
H to 5m (15ft); S 1.5m (5ft).

5 *Pseudosasa japonica* **Arrow bamboo**
Spreading bamboo that will form tall thickets in warm climates. Young canes are olive green, but pale beige when mature.
H to 6m (20ft); S indefinite.

6 *Phyllostachys nigra* **Black bamboo**
Canes become almost jet-black by their third year. A neat clump-former.
H 3–5m (10–15ft); S 2–3m (6–10ft).

7 *Chusquea culeou* **Chilean bamboo**
Distinctive, thick canes with white, papery sheaths in the first year, and prominent, swollen "knuckles" or nodes.
H to 6m (20ft); S 2.5m (8ft).

8 *Phyllostachys viridiglaucescens* **Phyllostachys**
Hardy, evergreen clump-forming bamboo with greenish-brown canes that arch at the base.
H 6–8m (20–25ft); S indefinite.

9 *Yushania maculata* **Yushania maculata**
Upright, spreading bamboo with narrow, green leaves; canes turn from blue-grey to olive green.
H to 4m (12ft); S indefinite.

10 *Fargesia murielae* **Umbrella bamboo** or **Muriel's bamboo**
Compact, elegant, arching bamboo with cascades of delicate leaves and canes that turn from green to yellow as they mature.
H 4m (12ft); S 1.5m (5ft).

11 *Sasa veitchii* **Veitch's bamboo**
Spreading bamboo with purple canes and broad, dark-green leaves that have white, parchment-like margins in autumn.
H 1.5m (5ft); S indefinite.

12 *Phyllostachys nigra* f. *henonis* **Henonis bamboo**
Bright green canes turn yellow-green when mature. Glossy, evergreen leaves, which are downy and rough when young.
H 10m (30ft) S 2–3m (6–10ft).

13 *Semiarundinaria fastuosa* var. *viridis* **Narihira bamboo**
Tall, upright, clump-forming bamboo with green canes – ideal for woodland gardens.
H to 7m (22ft); S 2m (6ft).

14 *Thamnocalamus spathiflorus* **Thamnocalamus**
A fast-growing, clump-forming bamboo. The stems turn an unusual pinkish-brown as they age.
H to 10m (30ft); S 6m (20ft).

15 *Fargesia murielae* 'Simba' **Umbrella bamboo 'Simba'**
So-called dwarf variety but may still grow to a good size. Compact and clump-forming evergreen.
H 1.9m (6ft) S 1.5m (5ft).

16 *Pleioblastus variegatus* **Dwarf white stripe bamboo**
Small, bushy, spreading bamboo with clearly striped, green and white leaves.
H 0.75–1.5m (2ft 6in–5ft); S 1.2m (4ft).

GROWING FERNS

Ferns are unlike all other plants. For a start, they don't flower – so their appeal is much more to do with form and texture than with dramatic differences in colour. They also reproduce completely differently. Instead of setting seed, most produce spores. These are contained in tiny capsules on the undersides of the leaves. When the spores are released, they may – if conditions are right – produce small growths that will eventually grow into new, young plants. It is possible to propagate ferns yourself, by collecting spores and growing them on in pots kept in a propagator, but it's hit-and-miss and can take almost two years before you have plants mature enough to plant out. It's easier to propagate them by division (*see pp.304–305*). Some ferns are hardy and will happily grow outdoors; others are tender and are best protected or moved indoors in winter.

Ferns planted in a mixed border look wonderful in spring as their tightly coiled new fronds gradually unfurl – although they will do much better if the bed is in at least partial shade.

Tender, sub-tropical tree ferns such as *Dicksonia antarctica* can be grown outdoors in sheltered, temperate gardens, but they are not hardy and their trunk or crown may need wrapping in fleece or straw to be sure they survive the winter.

A damp, north-facing garden can create a real challenge for growing anything except for ferns and other shade-tolerant, moisture-loving plants.

Where to plant ferns

Ferns are a godsend for any gardener struggling with a shady, damp area where few other plants are happy. Many ferns positively dislike full sun, since their fronds are prone to scorching in the heat. They much prefer humidity, and most will thrive in full or partial shade, in boggy corners, besides ponds and streams, and beneath the branches of overhanging trees. The most important things are to keep them moist at all times, to plant them in rich, fertile soil that has had plenty of well-rotted organic matter dug into it, and to protect them from wind. Ferns tend to be slow growing and may take some time to establish themselves. Thereafter, most are tough survivors and should require little more than the removal of dead foliage in spring when new fronds start to emerge.

Ferns tend not to be over-fussy about the pH level of the soil, and most will grow happily if it is neutral to alkaline. There are one or two exceptions, however: so-called hard ferns (*Blechnum*) prefer acid soils, as does the small, delicate parsley fern (*Cryptogramma crispa*). If you have naturally alkaline soil, it may be easier to grow these in containers filled with bark-based compost or special ericaceous (lime-free) compost mixed with sharp sand or grit.

Tree ferns can be grown outdoors but only in warm, sheltered sites. They are tender, sub-tropical plants originating from the South Pacific, Australasia, South and Central America, and South East Asia. In temperate northern climates, they must be brought indoors in winter, or the trunks from which their long fronds sprout must be wrapped in a covering of straw or fleece.

HARDY FERNS

1 *Dryopteris filix-mas* **Male fern**
Deciduous fern forming large, shuttlecock-like
clump of upright, arching, mid-green fronds.
H 1.2m (4ft); S 1m (3ft).

2 *Osmunda regalis* **Royal fern**
Large, deciduous fern with bright green fronds.
Fertile fronds have rust-brown, flower-like spires.
H 2m (6ft); S 4m (12ft).

3 *Matteuccia struthiopteris* **Ostrich fern** or
shuttlecock fern
Tall, sterile, green fronds up to 1.2m (4ft) in
length surrounding shorter, brown fertile
ones, arranged in the shape of a shuttlecock.
H 1.7m (5ft); S to 1m (3ft).

4 *Polystichum setiferum* **Soft shield fern**
Evergreen fern with arching, lance-shaped
fronds; will form large clumps in deep
shade with lots of moisture.
H 1.2m (4ft); S 90cm (36in).

5 *Asplenium scolopendrium* Crispum Group
Hart's tongue fern
Glossy green fronds with distinctive
wavy or undulating, frilled edges.
H 30–60cm (12–24in); S 30–60cm (12–24in).

6 *Polypodium vulgare* **Common polypody**
Low-growing, spreading fern with leathery,
dark green fronds. Good for ground
cover – and will tolerate sun.
H 30cm (12in); S indefinite.

7 *Dryopteris affinis* **Golden male fern**
Slightly shorter than the standard filix-mas,
with a dark spot where each pinna (part of
the frond) joins the golden-brown midrib.
H 90cm (36in); S 90cm (36in).

TENDER FERNS

1 *Platycerium bifurcatum* **Common staghorn fern**
Long, arching, lobe-shaped, antler-like
fronds give this fern its common name.
H 1m (3ft); S 1m (3ft).

2 *Nephrolepis exaltata* **Sword fern**
Tufted, evergreen or semi-evergreen
fern with long, widely arching fronds.
H to 2m (7ft); S to 2m (7ft).

3 *Pteris cretica* var. *albolineata* **Cretan brake**
Evergreen fern with attractive, variegated
fronds that have white central bands on
each individual pinna.
H 45cm (18in); S 30cm (12in).

4 *Adiantum raddianum* 'Gracillimum'
Maidenhair fern
Extremely delicate fern with finely divided,
almost cloud-like fronds. Needs warmth
and humidity.
H to 80cm (32in); S 30cm (12in).

5 *Asparagus densiflorus* 'Myersii' **Foxtail fern**
Erect, foxtail-like fronds of dense, feathery
stems, each 30–40cm (12–16in) long.
H to 1m (3ft); S 50cm (20in).

6 *Blechnum gibbum* **Hard fern**
Tall, evergreen fern with long, straight,
deeply divided, bright green fronds that
spring up from a trunk-like rhizome.
H to 90cm (36in); S to 90cm (36in).

LAWNS, MEADOWS & GRAVEL GARDENS

A lawn can have a magical effect on a garden. The open area of green it provides is soothing both to look at and to walk on. Lawns are very flexible, as they can be cut to any shape from a rectangle to sinuous curves that provide open areas and link seamlessly to other parts of the garden. Most gardens, even small ones, are likely to have a lawn. A traditional lawn should be an even green with a crisp edge; otherwise it is unlikely to enhance the beauty of your garden. Look at your lawn closely: does it pass muster, or is it scruffy around the edges and marked with brown patches? Is the grass fairly uniform, or are there assorted broad-leaved weeds, mossy or bare areas under trees, or coarse grasses sprouting in clumps?

A new turf lawn is an expensive but instant option; there is no waiting for seeds to germinate or young grass to establish itself.

WHAT SORT OF LAWN?

An attractive lawn does not have to be closely cut and evenly striped. Such an effect needs a lot of work, and the fine grasses that give the smooth appearance will not tolerate heavy wear. If your lawn is well used, coarser grasses cut a little higher will suit your purposes much better. Grass does not have to be cut to a uniform height: you could leave an area to grow long and flower. This looks appealing and attracts wildlife, especially if planted with wild flowers and bulbs.

SIZE AND SHAPE

It is easier to mow and edge a simple shape. If your lawn has awkward corners, you could cut them out or broaden tight curves. For small gardens, a geometric shape works best; rectangular, round, or oval lawns are ideal. On small lawns, neat edges are important and weeds are very noticeable. Large lawns are easier to manage in many ways. Weeds are less obvious, and wear and tear makes less impact. Ride-on mowers can be used, and gently curving shapes are easy to handle.

Steeply sloping banks can be mown with a robot mower, but a walled terrace or ground-cover planting may look better and be easier to manage than grass. Gently undulating lawns are not a problem, but if your lawn has humps that are scalped by mower blades, leaving bare patches, and hollows that fill with lush long grass, levelling the ground is advisable.

Grass paths are feasible where wear is light. Make them of mowable width, at least 90cm (36in); calculating the width in multiples of your mower blade will make mowing much easier. Remember the edges – are you prepared to trim them? Is there a hard edge? A hard edge of bricks, kerbstones, treated timber or metal strip set below the level of the grass allows the mower to pass over the edge freely, saving a great deal of work.

(opposite, clockwise from top left)
A neat patch of green for summer use describes most lawns. An even turf requires some attention, but a few weeds won't matter.

To encourage wildlife, cut the lawn only in late summer and late autumn. Sow wild flowers to create a mini-meadow.

For a playground, use a coarser lawn mix with rye grass. The grass should not be cut too short, and needs regular care.

A perfect green lawn to set off the garden needs a high proportion of fine grasses. Paths and summer watering are essential.

A grass path can be a great alternative to paving in low-traffic areas.

LOOKING AFTER YOUR LAWN

To keep a lawn dense, green, and springy you need to feed and weed it, rake off debris and dead grass, and every so often get air into the soil after the wear and tear of summer. However, mowing is the most important job and this must be done regularly while the grass is growing. Grass clippings don't need to be a waste product; they can either be used as a mulch on the grass to reduce evaporation, or can be added to the compost bin where they will break down.

(top) **Rotary mowers** are good for utility lawns and long grass; some have a mulching facility or roller to leave stripes.

(bottom) **Strimmers** are useful for long grass and where the lawn abuts a wall. On some, the head can be turned to trim edges vertically.

MOWING

Regular mowing encourages dense growth. Mow little and often, removing no more than a third of the grass blade at each cutting. If you routinely let the grass grow long and then cut it hard, the quality of the lawn will deteriorate.

Grass will grow fastest in warm, moist conditions, when a fine lawn may need cutting two or three times a week. A weekly cut is usually sufficient for a utility lawn; remember that coarse-leaved grasses do not tolerate close mowing. In hot, dry weather, mow less frequently and allow the grass to grow longer than normal in order to conserve moisture. Mix the clippings into the compost heap. If you plan to apply weedkiller, mow the grass tight beforehand. Afterwards, continue to mow frequently, leaving the clippings to fall into the sward – or, if this is not possible, leave them to rot in an out-of-the-way place.

Clean your mower after every use, removing any grass and wiping the blade with an oily rag. Instructions come with every machine, and it is worth taking note of these. Regularly oil moving parts, and keep the blade sharp or it will tear the grass. It is easy to remove and sharpen the blades of a rotary mower, using an oilstone or diamond sharpener or file. There are gadgets available to sharpen the blades of cylinder mowers, but these are not always very effective. It may be better to have the blades professionally reground. On some models they can easily be replaced when they become worn or damaged.

MOWING TIPS

- **Mow in spring,** summer, and autumn, and occasionally in winter if the grass is growing.
- **Do not mow** when the lawn is very wet, frozen, or during drought conditions in summer.
- **Consider using a mulching mower** for all but fine turf.
- **Brush off** any worm casts before mowing, or they will be flattened and smother the grass.
- **Collect clippings.** If left they will encourage earthworms and worm casts, and build up a layer of "thatch". The exception is during hot summers, when the clippings can be left as a mulch to conserve soil moisture.
- **For a rectangular lawn,** first mow a wide strip at each end, for turning the mower. Then mow up and down in straight strips, slightly overlapping the edges of each strip.
- **For a curved or irregular shape,** start by mowing a strip around the edge, then mow straight up the centre. Mow each half as for a rectangular lawn.
- **If a mower jams** while you're using it, make sure you unplug it or remove the spark plug lead before investigating.

Plants spilling out of borders shade out grass and make mowing awkward. Cut back sprawling stems or lay a hard paved edge to help keep borders neat.

WATERING

Grass loses its springiness and starts to discolour when it is short of water. New lawns and high-quality lawns must be watered, unless there is a hosepipe ban. Water as soon as you notice that the grass does not spring back after being walked on. Utility lawns can be left; the tougher grasses in these lawns will turn brown as they become dormant, but will recover and go green when rain arrives. If at all possible, it is easier and less wasteful to wait for rain.

Lawn sprinklers are the usual method of watering, but are potentially wasteful. To reduce evaporation, run them in the morning or evening, or during the night. You can also calibrate sprinklers so that they only apply enough water to soak the top 15cm (6in) of soil. For large, high-quality lawns, consider installing pop-up sprinklers or investing in a robotic watering device that moves around automatically after an area is soaked.

FEEDING

A starved lawn will turn a pale yellowish green. To avoid this, feed your lawn at least once a year with the lawn fertilizer appropriate to the season. Check that you are buying what you need and always follow the manufacturer's instructions.

- **Spring and summer feeds** are high in nitrogen to encourage rapid, lush growth and good colour. Apply in spring or early summer.
- **Autumn feeds** are high in potassium to toughen up the grass for winter. Apply in early autumn.

Lawn feeds come in granular and liquid formulations. Liquid feeds are watered on. They are fast-acting and most suitable for small lawns. The granular forms have a longer-lasting effect, releasing nutrients more slowly, which can limit the risk of scorch in summer. However, they need watering in if there is no rain within two or three days. They must be spread evenly, so invest in a wheeled spreader if your lawn is large; this will need to be adjusted so that the feed is applied at the correct rate.

AUTUMN LAWN CARE

The droughts and wear and tear of summer can leave lawns hard and patchy. Improve them by scratching out dead grass "thatch" and opening up compacted soil to allow the roots to breathe and rain and nutrients to penetrate. This helps grass to grow deeper and survive drought. Rake up leaves and brush off worm casts; where grass is covered for any time, it turns yellow and may die.

1 Use a spring-tined rake to vigorously scratch out or "scarify" thatch.

2 Aerate the lawn every 2–3 years. This is easiest when the soil is moist. Drive in a fork, or on better soils use a hollow-tiner to remove cores of soil, every 10cm (4in).

3 Apply top-dressing directly after aerating, spreading the mixture evenly over the surface – one bucketful will cover approximately 5 sq m (6 sq yd). Make top-dressing using 3 parts sandy loam, 6 parts horticultural sand, and 1 part peat substitute (such as composted bark or coir).

4 Work the top-dressing well into the grass and the air holes using the back of a rake or a stiff broom. To make the job easier, allow the mixture to dry slightly before brushing it in.

5 Apply an autumn lawn feed at the rate recommended on the packet. For an even spread, scatter half the fertilizer in one direction and the other half at right angles to it.

6 If the lawn looks sparse, sprinkle over some seed. Choose a seed type to match the rest of the lawn and use half the amount recommended for sowing a new lawn.

LAWN EDGE REPAIR

If not repaired, damaged edges will continue to deteriorate, and like any bare patch, they also provide an open invitation for weeds to establish. Reseeding a damaged edge is rarely effective, so try tackling the problem using this simple technique.

1 Cut out a rectangle of turf around the damaged area. Cut a generous area, or the turf will fall to pieces as you lift it out. It might help to stand on a board and cut against it to get a straight edge.

2 Undercut the rectangle with a spade. Try to cut the turf to as even a thickness as possible, with at least 2.5cm (1in) of soil.

3 Lightly fork over the newly exposed soil to break up the surface and to encourage re-rooting. Add soil if necessary and firm well.

4 Lay the piece of turf again, turning it round so that the broken edge faces inwards, and the cut edge aligns with the edge of the lawn. Check the fit and level, and make any adjustments. Butt up the edges, pressing down the refitted section firmly.

5 Fill the hole with top-dressing (see *pp.172-173*) and work it into the joins to help them knit together. You can reseed the hole if it is large, using an appropriate grass seed. Water, and keep an eye on the repair until the grass has started to regrow, weeding and watering if required.

6 To maintain a neat edge, recut it whenever it starts looking untidy; this is easiest when the soil is moist but not wet. Lay a plank on the lawn to stand on and cut against, or use a taut length of string as a guide. Sever the grass roots with a spade or edger where they are spreading into the bed or border. When you have finished working in one direction, work back in the other, pressing against the cut surface to firm it.

WHAT'S WRONG WITH MY LAWN?

All lawns inevitably develop discoloured patches or problems from time to time, often caused by adverse weather conditions. Keep an eye out for the first signs of trouble, because prompt treatment will usually prevent permanent damage to your lawn. If the same problems keep reappearing time and again, look at the underlying conditions in your garden; tackling the causes will be far more effective in the long run than repeatedly treating the problems they create.

LACK OF LIGHT

Grass growing in light or dappled shade can thrive, but the deep shade from trees can often be too much. First of all the grass will become thin, and then moss is likely to develop. It may be possible to open up the tree canopy by carrying out judicious pruning in order to let more light penetrate. Shade-tolerant seed mixes are available, but they are fine grasses that do not cope well with heavy wear or close cutting. Consider using alternative groundcovers in shady areas, such as moss, shade-loving plants, or a bark mulch.

WATERLOGGING

Poorly drained lawns are beset with problems as grass roots die through lack of oxygen. If water sits on your lawn after rain, it can point to a high water table or to poor drainage.

An impervious layer of compacted soil – known as "hardpan" – below the surface can often result in poor drainage. It is possible to relieve compaction and penetrate the hardpan by spiking if it is not too far below the surface (see *pp.172–173*); smeared layers of hardpan near the surface are common.

Scorch/yellow patches may be due to fertilizer that was unevenly distributed or not watered in, or to dog urine. If a dog urinates on the lawn, water the area thoroughly.

Snow mould is most common in late autumn and winter. It is a sign of poor aeration and too much nitrogen. Scarify and aerate the lawn, and do not use spring lawn feeds in autumn.

Drought causes yellowish, straw-like patches; the whole lawn may be affected. Let the grass grow longer and leave clippings on the lawn to reduce evaporation. Water or wait for rain.

If the problem lies deeper, more drastic and expensive action will probably be required. Solving the problem may either involve digging drainage trenches or laying a system of drainage pipes.

Of course, it could be easier to abandon all thoughts of establishing a lawn in these circumstances, and instead grow a variety of plants that thrive in moist conditions (see *pp.50–53*).

Lawns on a clay soil can often be especially problematic, as winter wet leads to dead patches and moss growth, while in summer the roots of the grass may fail to reach moisture deep in the soil. One way of overcoming this is to copy the method of making golf greens, and lay the lawn on a layer of coarse sand.

Start by improving the soil using generous additions of organic matter as well as some careful cultivation. Then spread a layer of sharp sand 5–8cm (2–3in) deep over the surface and place the turf on top. The grass should root through the sand into the fertile clay below and make good growth. During summer the roots will be able to extract moisture from the moist soil below the sand, and in winter excess water will drain through the sand, which should result in a lawn surface that is far less wet.

WEEDS

Rare is the lawn without a weed; weed seeds are either blown into your garden or brought in by birds. A good lawn-care regime will deal with many weeds: regular mowing exhausts them, and healthy, strong-growing grass will crowd them out. Mowing the lawn too short, letting it grow too long, or failing to feed can all provide an opening. Weeds will move into bare patches where the grass has been worn away or weeds have been dug out by hand; prevent this by reseeding. If you cannot match the seed mix to the existing lawn, it will soon blend in.

Rake the lawn before mowing to lift up weed stems into the path of the blade, and always use a grass box to prevent weeds spreading. For difficult weeds, use a suitable lawn weedkiller, or apply a weed-and-feed formulation in spring.

Moss in a lawn is a sign of bad growing conditions, such as low fertility, compacted or poorly drained soil, and shade; it is also a problem on acid soils. Autumn lawn care and spring feeding will help (see *pp.171–173*). Test the soil, and if the pH is low, apply lime in winter in the form of ground lime at the rate of 50g per sq m (1½oz per sq yd).

Worm casts, left by earthworms, are small, muddy mounds on the lawn in spring or autumn. These are unpleasant, but not damaging, and are nutrient-rich. Brush them into borders before mowing.

Fairy rings appear as lush grass in spring. Underneath are fungi living on matter like old tree roots, and toadstools appear in autumn. Feed, aerate, and water the lawn to keep them in check.

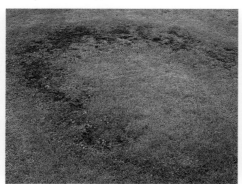

Mole hills are unsightly. Ultrasonic devices or smoke in the tunnels may drive the moles away, but trapping and releasing them elsewhere is more likely to be effective.

LAWN WEEDS

Typical lawn weeds fall into two types: low-growing wild flowers, and undesirable coarse grasses or grass-like plants. They can all survive regular mowing. Once they have germinated, they spread by creeping stems, runners, or by seed in grass clippings. Dandelions have deep tap roots and if their tops are cut off, more small plants grow from any part of the root left in the soil. Daisies do the same, but to many they are a delight and not a weed. Some weeds may be encouraged in wildlife-friendly long grass; others smother lawn grasses and need controlling quickly.

1 White clover (*Trifolium repens*) Mat-forming weed indicating poor, slightly alkaline soil; its roots contribute nitrogen. Use a lawn weedkiller.

2 Annual meadow grass (*Poa annua*) This is a tufted grass that spreads by seeding; dig out plants by hand before they flower.

3 Creeping buttercup (*Ranunculus repens*) The creeping stems form mats. Weedkiller is not reliable; you may have to dig this out by hand, so try to catch it early.

4 Moss This indicates poor conditions, especially on acid soils. Use moss killer in spring or autumn, but it is likely to return if the underlying cause is unresolved.

5 Self-heal (*Prunella vulgaris*) A mat-forming weed, which is best dealt with by use of a lawn weedkiller. This could be left in meadow grass.

6 Slender speedwell (*Veronica filiformis*) Control the mats of tiny leaves by feeding the lawn to encourage growth and scarifying in autumn.

7 Dandelion (*Taraxacum officinale*) Use a spot weedkiller or dig out the tap root. Deal with dandelions before they have the chance to seed.

8 Pearlwort (*Sagina procumbens*) This is a creeping weed that forms low mats of grassy leaves. It can be controlled with a lawn weedkiller.

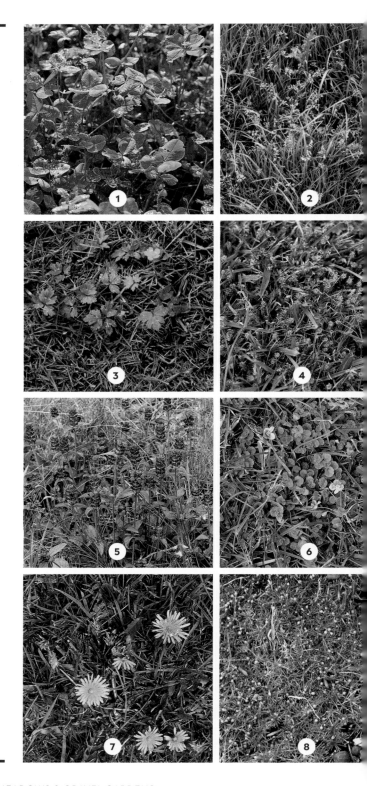

9 Lesser yellow trefoil (*Trifolium dubium*) An annual weed that is often quickly spread by mowing without a grass box. Control with lawn weedkiller.

10 Field woodrush (*Luzula campestris*) A clump-forming, grass-like plant which prefers acid soil. Dig out by hand; apply lime in winter in the form of ground lime.

11 Yorkshire fog (*Holcus lanatus*) A coarse-leaved, creeping grass; slash repeatedly with a knife to weaken bad patches, or dig out and re-seed.

12 Broad-leaved plantain (*Plantago major*) A solitary plant with a tap root, which spreads by seed. Dig out the roots by hand or use a spot weedkiller.

13 Daisy (*Bellis perennis*) Leave the rosettes if you like a daisy-spangled lawn. If not, dig out the roots by hand.

14 Mind-your-own-business (*Soleirolia soleirolii*) Forms mats of tiny leaves, thriving in damp sites. Lawn weedkillers do not effectively suppress it; rake it out.

15 Yarrow (*Achillea millefolium*) The mats of ferny leaves thrive in dry soil; they will also stay green in dry weather. Dig it out by hand or use lawn weedkiller.

MAKING A NEW LAWN

If you want a new lawn, you have the choice of laying turf or sowing seed. Whichever method you choose, the site must be well drained, level, and cleared of weeds. This means preparing the soil thoroughly, at least five weeks and ideally several months in advance, to give it time to settle.

PREPARING THE SOIL

Dig over the area, plus an extra 15cm (6in) all round, removing weeds and adding soil improvers (see *pp.172–173*). On heavy clay or wet soils, mix sand into the topsoil to improve drainage (see *p.147*). On light, sandy soil add 8–10cm (3–4in) of bulky organic matter during digging; this will help to retain moisture and nutrients.

1 After digging and levelling, while soil is moist but not wet, use the back of a fork to break down the soil surface until it is fine and crumbly, then rake level. Tread over the site with the weight on your heels to eliminate soft spots that might later sink, giving an uneven surface. Repeat at right angles.

2 Rake off stones and debris; if soil is stony, rake again at right angles. After leaving for at least five weeks to allow soil to settle, lightly hoe off any seedling weeds. A few days before laying turf or sowing, lightly rake in a balanced organic or compound granular fertilizer at the recommended rate.

CREATING THE RIGHT ENVIRONMENT

• Kill or clear all perennial weeds, plus the old grass if you are replacing a badly neglected lawn. The easiest way is to use a glyphosate weedkiller, although more than one application may be needed.
• Cultivate the soil, first skimming off any existing turf. If you prefer not to use a weedkiller, you can dig out weeds by hand during cultivation.
• A good lawn requires 20–30cm (8–12in) of well-drained topsoil; the subsoil should also drain well. With these conditions the grass will root deeply, reducing the need for watering. Where the depth of topsoil is irregular, the lawn will dry out more quickly in shallow areas and develop brown patches. If your topsoil is shallow, it is worth buying enough to achieve an even depth.

SEED OR TURF?

• **Seed** costs less than turf, and the wide choice of grasses means you can satisfy your requirements more exactly, but it will be up to a year before the lawn can take heavy use.
• **Turf** gives you an instant lawn, usable in about eight weeks, when the turf has taken root. However, good-quality turf, like good-quality carpet, is not cheap.
  Buy turf from a reputable specialist. Turf from good suppliers is purpose-grown in fields and lifted to order. If you can inspect the turf to check the quality, so much the better. Usually there is a choice between high-quality ornamental turf, containing only fine grasses, and utility grade, which has a proportion of hardwearing rye grasses. Some suppliers will grow special turf to order, but this is more expensive and takes about 18 months.

LAYING A TURF LAWN

The best time to lay turf is early autumn or early spring. Turf should not be left lying about, so set to work immediately; otherwise unroll each turf and lay grass side up on paving or plastic sheeting, watering if necessary. It will keep for a few days like this.

Turf should be laid on soil that is moist, but not wet. Lay the first row of turves along one edge of the site, tamping down each piece with the back of a garden rake. This will eliminate air pockets and bring the turf into close contact with the soil, encouraging it to root.

Keep the lawn well watered, or turves will shrink and curl. In dry weather, water up to twice a week at first – this encourages rooting.

Do not disturb the turf before it has rooted; try to keep off it for at least three weeks. To check rooting, try peeling back one corner of the turves; if they do not lift easily, the grass has rooted. Once this has happened, you can mow the lawn as necessary but set the blade high. Avoid all but the lightest use until the lawn has been down for eight weeks.

1 Make a tight seam with the adjacent piece of turf. To do this, raise the edges slightly, push them together so they are almost overlapping, and press down firmly with your thumbs. Tamp down once more along the seams to prevent the edges from lifting.

2 Continue to lay the turves in rows, arranged so the joints are staggered, like bricks in a wall. Protect the turf that has already been laid by working on a plank to lay the next row. Continue until the site is covered, firming by tamping down with the back of a rake. Do not use small pieces of turf at the edges; they will quickly dry out. Use a full-size turf at the edge, and put the smaller piece in the gap behind it.

3 Scatter fine sandy loam, or good topsoil mixed with horticultural sand, over the surface. Brush it well into the joins to fill any gaps and encourage the turves to knit together.

SEEDING A LAWN

The best time to sow a lawn is in early autumn when the soil is warm and moist. The next best time is spring, but the soil may be cold after winter, which will delay germination. Scatter seed by hand; a seed distributor may be useful to cover large areas. Sow the seed over a slightly greater area and cut a clean edge a year after sowing when the lawn is established.

1 Prepare the ground for the seeds. If hand sowing, divide the site into 1m (3ft) squares with string or canes. Stir the seed to mix evenly. Weigh out seed for 1 sq m (1 sq yd), following the directions on the pack. Pour into a plastic cup, mark the level and use it as your measure. As a guide, 30g (1oz) is roughly a handful.

2 For even distribution, scatter half the seed over the square in one direction, and half at right angles to it. Use the same criss-cross method if sowing by machine. Repeat for each square. When all the seed is sown, remove the strings or canes.

3 Seedlings should appear within 14 days. Water regularly. Make the first cut, on a high setting, when seedlings have reached 5cm (2in). For autumn-sown lawns, maintain this height until growth slows. The following spring you can gradually lower the blade height.

ALTERNATIVES TO GRASS

If you decide your lawn is too small, too rough and weedy, too steep,
or too much trouble, there are some practical, low-maintenance alternatives.
Choose soft play bark for a children's play space, pave it for an all-weather
entertaining area with raised beds, or create a sunny gravel garden.

GROUND-COVER OPTIONS

In a sunny, well-drained garden, spread gravel 2.5cm (1in) deep over a geotextile membrane; this will suppress weeds and create a seaside mood. If you plan to lay down paving, it needn't stay dull and grey. Leave out a slab here and there to make room for a small tree or low shrub, and plant small ferns or creeping Jenny in moist, shady cracks.

Plants such as cyclamen, hellebores, or Vinca minor grow well under trees. Some, like lilyturf (*Liriope muscari*) and ajuga, even put up with dry soil. Once they have settled they will look after themselves, especially if you lay a mulch to suppress weeds and retain soil moisture. While grass might not tolerate a very shady, damp area, this is the perfect environment for moss to establish; let it spread, don't walk on it too often, and plant early-flowering bulbs like snowdrops (*Galanthus*) or lily-of-the-valley (*Convallaria majalis*).

When laying down paving, plant creeping mints and thymes in sunny crevices and gaps, where they will release scent when trodden on.

(top) **Cyclamen** are quite happy growing in the shade cast by trees, providing ground cover and brightening up a shady spot.

(bottom) **Moss** often grows naturally in damp shade where grass struggles.

WILDFLOWER MEADOWS

A meadow is an area where grasses and flowering annuals, bulbs,
and perennials are allowed to intermingle and seed themselves – it looks
natural and artless. In reality, getting the balance right can be quite difficult,
especially in a traditional meadow of native species of grasses and flowers. You do
not have to use native plants, though. The term "meadow gardening" includes a
variety of naturalistic planting styles and you may find that it is easier to create
a meadow with a range of annual species or a mixture of native and non-native
perennial cultivars and grasses. As well as being beautiful, an established meadow
is also a wildlife haven, providing not only shelter, but nectar and seeds to sustain
a varied colony of local fauna. Even a small meadow encourages biodiversity by
providing a "green corridor" that enables the wildlife to travel across the landscape.

CREATING A MEADOW

Planning is crucial, as the plant community will need
to develop with little interference on your part.
Assess the soil type and growing conditions of the
site, then choose plants appropriate for the plot. If
you don't, not all the plants in the meadow will thrive
equally, causing weaker species to disappear and
stronger ones to take over. Contrary to popular belief,
meadows can occur both on free-draining soil and on
damp ground. Meadow-seed specialists offer a range
of mixes to suit different soil types and conditions.

(top left) **Annual meadows** create a glorious show of flowers;
the cornfield annuals must like freshly cultivated soil, and have
to be prolific self-seeders, since the meadow is dug over each
year. Here, scarlet poppies, blue cornflowers, and white ox-eye
daisies predominate.

(bottom left) **A traditional flowering meadow** such as this one
consists largely of native species, which have smaller, more
delicate flowers than the cultivated varieties that you usually
find within typical garden borders.

• **Traditional wildflower meadows** depend on low soil fertility. If the soil is fertile, the wildflowers will not be able to compete with vigorous grasses. After a few years, few flowering plants will remain. Therefore choose an area that hasn't been cultivated recently. You can either sow a meadow in a bare piece of ground, or you could plant wildflowers in an established grass patch.

• **Annual meadows** spring up when land is left after the soil is disturbed; seeds in the soil quickly germinate. First come annuals like poppies and cornflowers, before perennial grasses and flowers take over in following years. To prevent perennials from establishing, dig over a meadow each year, leaving the annuals space to spring up anew. To sow the annual meadow, either use a prepared mix, or buy your own choice of annual seeds and mix them for a natural, random effect. Sow in spring, or in autumn if your mix consists of hardy annuals. Turn the ground in autumn, and the display will be reproduced the next year.

• **Vibrant meadows** can be created by purposely supplementing the subtle delicacy and prettiness of native wildflowers. To produce a bolder display, include non-native perennials and their cultivars, which generally have larger, brightly coloured blooms. You could also add bulbs such as crocuses, daffodils, tulips, and snowdrops.

Preparing to sow

You can slightly reduce fertility in a lawn in the year or two before sowing a meadow by mowing repeatedly and removing the cuttings, which prevents nutrients returning to the soil. Stripping off the topsoil (the most fertile layer) also helps, but it is hard work and you'll need to find a new home for it. If you have a deep, rich topsoil, it may not even be possible to remove it all. Try to prepare the soil a few weeks before sowing (*see right*). This allows any annual weed seeds in the soil to germinate, so that you can hoe off the seedlings before sowing.

SOWING A WILDFLOWER MEADOW

Sow in autumn, as some dormant seeds won't germinate unless they experience a period of cold first; some may not germinate until the second year. You'll need about 15g seed to 1 sq m of soil (½oz to 1 sq yd), plus a bulking agent such as dry sand.

1 Mix the sand and seed, and divide the volume by the number of sq m (or sq ft) of your plot, so you know how much you'll need to sow in each section.

2 Dig over the area to ensure it is not compacted, removing any weeds. Level it with a rake, removing large stones, and tread over the site to firm. Rake it lightly to form a crumbly surface. Divide the plot into squares, and sow one square at a time to ensure even coverage.

3 Cover the seed thinly with soil. Firm by tamping down with a rake, or by lightly treading or rolling the plot.

GRAVEL GARDENS

While lawns require frequent mowing, as well as intensive care at
different times of the year, gravel gardens need little attention and
look neat and attractive all year. A gravel garden also offers the ideal habitat
for plants that originate in warm, dry climates. These include those with grey and
silver foliage, such as artemisias and lavenders, which like the free-draining, warmer
conditions in a gravel garden. Surrounding warm-climate plants with a "collar" of
gravel helps them in several ways. Gravel reflects light, making optimum use of the
available sun. It also improves drainage around the plant bases, preventing delicate
or young plants from rotting off. When it rains, the gravel stops mud from splashing
onto foliage, which helps plants with hairy or woolly leaves to survive the wet winter
months – if they become too sodden and muddy, they are likely to rot.

PLANNING A GRAVEL GARDEN

Gravel gardens can look either formal or
naturalistic, depending not only on the way you
use gravel with other hard-landscaping materials
or structures, such as water features, but also
on how you plant it. The gravel could simply fill
in the spaces between the defined lines of formal
flowerbeds or pathways, or, for a more informal
look, you could use it to cover the entire garden,
with planting inserted into it in irregular drifts.
Alternatively, you could use several types of gravel,
or paddlestones, such as slate, to create interesting
patterns or contrasts of colour and texture.

If you garden on a heavy soil, and want to have
a gravel garden, the first, and crucial, step is to
improve the drainage. Incorporating coarse sand
and gravel into the soil will do this, but you need
to add a substantial amount, not just a scattering.
It is hard work, but worth it.

Lush planting is still possible in gravel gardens. Here,
purple-flowered nepeta, or catmint, has been planted
en masse to create a striking design.

Paving and gravel on this sheltered terrace provide planting pockets for sun-loving thymes. These spread out to form a fragrant carpet, and combine with geraniums and *Alchemilla mollis* in a low-maintenance plot.

PREPARING THE PLANTING SITE

Invest some time in preparing the area for your gravel garden, and you'll only need to spend a few hours each season maintaining it. First, clear the proposed site of all weeds and vegetation, and then condition and level the soil (see *pp.180–181*).

It is always wise to use a weed-suppressing membrane in a gravel garden. These are available in sheets, and are made either from a variety of materials woven into a fabric, or from perforated plastic. The membrane forms a weed barrier over the soil, eliminating light needed for seeds beneath to germinate and grow, but allowing rain to penetrate to ornamental plant roots. If weed seeds germinate on top of the membrane, they will not be able to root through it, and are easy to pull out.

Make sure that the membrane overlaps neatly, and peg it down with a few metal pins. Arrange your plants in their pots on top of the membrane to judge the overall effect and check spacings. Avoid using plants that naturally spread by sending up shoots from the roots. When you are happy with your design, plant through the membrane, and spread a thick layer of gravel over the surface. Alternatives to gravel include crushed stone, slate chips or paddlestones, cobbles, or glass beads.

A layer of organic matter will eventually form between the membrane and gravel. A few weeds may root into it, but not deeply, so they will be easy to remove by hand weeding the area a few times a year. Or you can use a flat hoe, sliding it between the membrane and gravel to dislodge the weeds' roots. On a dry day, the weeds will soon shrivel up.

GRAVEL GARDEN PLANTS

Choose drought-tolerant plants that will appreciate the free-draining conditions. These plants usually have foliage that is adapted to cut down on water loss – look for leaves that are silver in colour, narrow and strappy, or covered in fine, downy hairs. Many spring- or autumn-flowering bulbs tolerate dry conditions by becoming dormant during summer.

- *Achillea filipendulina* 'Gold Plate'
- *Alchemilla mollis*
- *Artemisia*
- *Eryngium giganteum*
- *Euphorbia characias*
- *Lavandula angustifolia*
- *Papaver* (Oriental Group) (see *below*)
- *Sedum spectabile*

MAKING A GRAVEL GARDEN

You could plant directly into the soil, if you wish, but you'll get better results if you lay a weed-suppressing membrane down first. Once you cover it with a thick layer of pea gravel or slate chips, you won't know it's there, but you will notice how it cuts down on weeding and watering. The membrane lets rain through, but like a super-efficient mulch, stops it evaporating.

1 Prepare the area by digging over the soil and removing any weeds. Dig in plenty of well-rotted garden compost or manure and, on heavy soil, add coarse sand or grit to improve drainage.

2 Cut a piece of weed-suppressing membrane to fit the bed or border. For larger areas, you'll have to join several strips together. Leave a generous overlap along each edge and pin securely in place.

3 Position your plants on top of the membrane. (Presoak container-grown plants in water for at least an hour.) Check that you have left enough room for each plant to spread to its full width. Using sharp scissors, cut a large cross in the membrane directly under each plant and fold back the flaps. Make it big enough to allow you to dig a good-sized planting hole.

4 Take each plant out of its pot and lower it into the hole: it should sit at the same level as it was in its pot. Fill in around the rootball with soil, then firm with your heel or hand.

5 Tuck the membrane flaps back around the plant stem. Pin the membrane in place if needed to close any gaps, taking care not to damage the rootball. Water thoroughly.

6 Spread a layer of gravel mulch, about 5–8cm (2–3in) thick, over the membrane. Be generous, to prevent the membrane showing through, and level with a rake. Water around the plants regularly until they are established, using a spray from a watering can or hose to avoid dislodging pockets of gravel. If any weeds get through, they should be easy to pull out.

PLANTS FOR GRAVEL GARDENS

1 *Lavandula angustifolia* 'Twickel Purple' **Lavender**
A mid-sized lavender that, like all others, thrives in well-drained soil in full sun.
H to 60cm (24in); S 1m (3ft).

2 *Dierama pulcherrimum* **Angel's fishing rod**
A summer-flowering plant that forms chains of corms each year. Once settled in, it's usually trouble-free.
H to 1–1.5m (3–5ft); S 60cm (24in).

3 *Deschampsia cespitosa* 'Goldtau' **Tufted hair grass**
Tussock-forming, evergreen grass producing silvery reddish-brown spikelets from early to late summer.
H to 75cm (30in); S to 75cm (30in).

4 *Sisyrinchium striatum* 'Aunt May' **Sisyrinchium**
A tall, clumping perennial with evergreen, striped leaves. It flowers in early to midsummer.
H 50cm (20in); S 25cm (10in).

5 *Artemisia alba* 'Canescens' **Wormwood**
Most artemisias thrive in gravel gardens. This variety has curling, silvery leaves.
H 45cm (18in); S 30cm (12in).

6 *Iris unguicularis* 'Mary Barnard' **Iris**
Beardless iris with tough, grass-like leaves and bright violet flowers in midwinter.
H 30cm (12in).

7 *Pennisetum alopecuroides* 'Hameln'
Fountain grass
A compact, early-flowering form, with greenish white spikelets, maturing to grey-brown.
H 0.6–1.5m (2–5ft); S 0.6–1.2m (2–4ft).

8 *Cistus* x *dansereaui* 'Decumbens' **Rock rose**
A low, spreading shrub with delicate, paper-thin, white flowers that last for only a single day.
H 60cm (24in); S 1m (3ft).

9 *Echinops ritro* 'Veitch's Blue' **Globe thistle**
Beautiful, rounded, dark blue flowerheads. A plant
that positively likes poor, dry soil in full sun.
H to 90cm (36in); S 45cm (18in).

10 *Hylotelephium spectabile* 'Brilliant' **Stonecrop**
Its large, persistent flowerheads appear in late
summer and attract butterflies and bees.
H 45cm (18in); S 45cm (18in).

11 *Eryngium giganteum* **Sea holly**
A tall but short-lived perennial, worth growing
for its persistent, thistle-like flowers.
H 90cm (36in); S 45cm (18in).

12 *Achillea filipendulina* 'Gold Plate' **Yarrow**
A clump-forming perennial with flat, golden-yellow
flowerheads that appear from summer to autumn.
H 1.2m (4ft); S 45cm (18in).

13 *Allium karataviense* **Allium**
A small allium that produces round heads of tiny,
star-shaped, pale pink flowers in summer.
H 10–25cm (4–10in); S 10cm (4in).

14 *Salvia officinalis* 'Tricolor' **Sage**
A variegated sage with aromatic, grey-green and
cream leaves flushed with pink or purple.
H to 60–80cm (24–32in); S to 1m (3ft).

15 *Alchemilla mollis* **Lady's mantle**
Drought-resistant perennial with downy leaves
on which rain and dew collect as attractive
water droplets.
H to 60cm (24in); S 75cm (30in).

16 *Dianthus* 'Little Jock' **Pink**
One of many pinks that bear long-lasting, sometimes
scented flowers all through summer.
H 20cm (8in); S 30cm (12in).

TREES

Whether your plot is large or small, consider planting a tree or two. Trees offer many benefits, sending the focus up to a wide, open sky in a small garden, and creating intimate enclosures in larger spaces. They also make wonderful focal points, particularly if you select a striking specimen with a weeping habit, attractive stems, decorative foliage, or colourful blossom and fruit. A handsome white-stemmed birch, multi-stemmed amelanchier, or fountain-shaped Japanese maple will draw the eye when planted at the end of a path, in the centre of a lawn, or against a wall or a smooth evergreen hedge. In a deep, rectangular plot, a tree can be planted half to three-quarters of the way down the garden to partially mask the view beyond, and bring the eye back to the foreground.

The paper-bark maple (*Acer griseum*) is a garden tree grown almost as much for its wonderful, peeling, orange-brown bark as it is for its foliage, which turns orange and scarlet in autumn.

GARDEN TREES

Trees provide permanent features in a garden. They will be there for a long time, and, once established, are difficult to move; think carefully about what you want to plant and where. Consider also the final height of the tree or, at least, its height after eight or ten years if it is a slow-maturing type. In small town gardens there is rarely space to plant a large tree; it will not only plunge your own house and garden into darkness, but the neighbours may suffer too. Don't be tempted by melancholic weeping willows or blue Atlas cedars. By the time they are too large for the garden, you will need to apply for felling permission from the local authority, and pay a tree surgeon a handsome sum to remove them without destroying anybody's property. Instead, think small, or choose a tree that may be pruned regularly to keep it in check.

POSITIONING TREES

A tree in the garden is like a large piece of furniture in a house: its sheer volume limits its placing. Before committing to a final position, ask friends to help by holding up long canes or broomsticks where you are planning to plant. Look at them from key points in the house, such as a kitchen or living room, and in the garden, perhaps from an entrance or seating area.

Consider, too, the position of the sun in relation to the tree and the shade it will cast. Think of how and when you use particular areas. To create a brightly lit breakfast area, for example, plant your tree on the west side of the garden, or position a tree on the east or north side of your plot to catch the setting sun on an evening drinks terrace. A play area needs shade during the hottest part of the day, while vegetable beds need sun all day.

A tree with colourful foliage, such as this variegated form of *Cornus controversa*, creates a spectacular show in a border, providing height, design focus, and a shady area beneath for woodland plants.

Trees also make beautiful features in a garden room. Several of the dogwoods, such as the early summer-flowering *Cornus controversa* or *C. kousa*, and Japanese maples (*Acer palmatum*), with their cut foliage and autumn tints, make striking focal points in a lawn, courtyard area, gravel garden, or at the end of a vista. Alternatively, select a tree with multiple stems such as a silver birch; the white trunks and branches take on a leading role in dark winter months when the leaves have fallen.

KEY FEATURES

Choose trees that will be happy with the soil and climate in your garden. A local nursery should be able to advise you as to what will perform best in your area. There are also a host of decorative features to take into account.

As trees flower for a comparatively short time, it may be best to regard blossom as an added bonus rather than the main attraction. Fruits, berries, or interesting seedpods are worth considering, as they often last longer than flowers. But the most enduring feature of any tree is the colour, texture, and shape of the foliage and stems. Feathery acacias and cut-leaf maples, for example, offer a light, soft touch, creating dappled shade. Use the layered open habit of *Cornus alternifolia* to introduce horizontal lines, or a cherry with an upright habit to add a vertical dimension. While autumn colour is a seasonal attraction, use trees with year-round yellow, red, or variegated foliage sparingly. The dark purple leaves of *Acer palmatum* 'Atropurpureum', *Prunus cerasifera* 'Nigra', or bronze-leaved *Malus* 'Profusion' can add drama but, to prevent them looking gloomy, set them against lighter colours. Too much yellow can also be overpowering, though trees such as *Acer japonica* 'Aurea', or *Liriodendron tulipifera* 'Aureomarginata', can brighten up dull areas.

(top) **The smooth, russet-coloured bark** of this cherry, *Prunus serrula*, peels off to give visual and tactile interest all year. The tree is also covered with white spring blossom, and in autumn the foliage turns a lovely yellow.

(bottom) **The weeping silver-leaved pear,** *Pyrus salicifolia* 'Pendula', makes a handsome feature in a lawn, or to punctuate a path in a small garden.

Double staking will give a newly planted tree extra support in a wind-buffeted position.

Rabbit guards will need to be in place for several years to protect young trees from damage.

PLANTING TIPS

Small trees establish themselves in a new spot much more easily than large ones do. The best pot size is one that you can carry home in a carrier bag. Keep the roots of bare-rooted trees shaded and protected from wind, preferably covered by a sheet of plastic or plastic bag, or under a mound of compost, to prevent them from drying out. Before planting, take a close look at the tree's roots and remove any areas that are damaged.

Place a length of wood or a cane across the centre of the hole to get the correct level for planting. Pot-grown trees should sit at the same depth as they were in the container. When a bare-rooted tree is held at the correct level, the point where the roots start to flare out from the trunk should be slightly proud of the soil. Alternatively, simply plant to the mark of the original soil level on the trunk. If the hole is too deep, remove the tree and put some soil back into the hole.

It is not necessary to enrich the soil unless you have very sandy or poor soil. In fact, this is likely to hinder establishment, encouraging the roots of the tree to remain within a nutrient-rich "comfort zone" at the planting site, rather than questing outwards and downwards in search of resources.

STAKES AND ANIMAL GUARDS

A single, low stake is generally agreed to be best for most young trees, allowing the stem to flex and strengthen in the wind without tugging at and dislodging the roots. However, for top-heavy and especially top-grafted trees (where a stem of one tree variety is grafted onto the main branch of another), such as the weeping Kilmarnock willow, a taller stake is advisable, and in very windy sites, two stakes can be used. You can use the stake to anchor netting guards in areas where rabbits and deer are a nuisance.

PLANTING A CONTAINER-GROWN TREE

Prepare your site by clearing it of weeds and, if you have very sandy or infertile soil, digging in organic matter over a wide area. Dig a hole, not too deep, but two to three times as wide as the rootball, and fork over and loosen the soil at the bottom. Avoid planting too deeply and in too narrow a hole – 90 per cent of all young tree deaths can be attributed to this problem.

1 Water the plant in its pot by plunging it into a bucket of water about an hour before planting. After this, remove it from its pot so that you can tease out any roots that have become compacted. Use a cane to establish the correct depth for planting; ensure that the plant will sit in the hole at the same depth as it was in its container.

2 Drive a stake low into the soil on the windward site of the rootball, and rotate the tree until the trunk sits comfortably parallel with the stake. A low stake allows the stem to flex, which encourages strong trunk growth.

3 Refill the hole with soil around the roots and the stake, and firm using your foot. Secure the tree to the stake using a buckle-and-spacer tree tie; this will need checking and loosening regularly as the tree grows. Water, and mulch as for a bare-rooted tree (see *pp.198–199*).

PLANTING A BARE-ROOT TREE

About four weeks before planting, prepare the ground. First, clear the site of weeds and grass: dig them out or use a glyphosate weedkiller. Dig in well-rotted manure or garden compost on infertile or sandy soils. Just before planting, remove any new weeds that have germinated. Water your bare-root tree by dunking it in a bucket of water about an hour before planting.

1 Dig a circular hole as for container-grown trees (see *p.197*). Lay a lathe or cane across the hole and use it to check that the dark soil mark on the stem of the tree is just above it.

2 Hold the tree upright – or, better, get someone else to hold it – and backfill the hole with soil, a little at a time, firming it in and around the roots with your fingers.

3 Once the hole is filled, start firming with your heel from the edge to reduce the risk of the soil sinking and the planting level becoming too low.

4 Water in well, then check the level again and fork over to fill any dips. There is no need to scatter fertilizer. In fact, this is more likely to cause harm by burning the roots, which will then not be capable of taking up nutrients for at least the first year.

5 Knock a short stake into the ground at an angle of 45 degrees. Secure the trunk to the stake using a buckle-and-spacer tree tie that allows for the stem to increase in girth without chafing. Check it at least twice during the growing season as the tree matures, and loosen if it becomes too tight.

6 Mulch around the tree using a fibrous, open-textured material, such as well-rotted compost. Spread it in a layer 5–6cm (2½–3in) deep, but do not let it come into contact with the tree stem, where the dampness may cause rot.

CARING FOR TREES

Dehydration is the most common reason for the poor growth
or death of young trees, so it is essential to keep your tree well watered
during dry spells from spring to autumn for the first two to three years after
planting. Give each tree at least 50 litres (11 gallons) of water per week; this equates
to about seven or eight full watering cans. If you're unable to water your trees
regularly while they establish themselves, try using a slow-release watering bag.
This drips water into the ground slowly and steadily, fully saturating the tree roots
over the course of the day.

MULCHING

To reduce competition for water and nutrients from grass, other plants, and weeds, keep clear an area of 1 sq m (1 sq yd) around the tree and mulch it every autumn, keeping the mulch away from the trunk. Young trees also benefit from an annual application, in spring, of a general fertilizer, such as blood, fish, and bone. Before you mulch, or after you have pulled back an existing mulch, scatter the fertilizer around the base of the tree at a rate of 70g per sq m (2oz per sq yd), fork it into the soil, then water. Once the tree is well established, you can let the grass grow around the trunk, or plant some bulbs or shade-loving groundcover underneath, without checking the tree's growth.

PRUNING YOUNG TREES

Conifers and other evergreens, such as hollies, seldom need pruning after planting, and some deciduous trees, including birches and amelanchiers, are best left alone to develop their naturally beautiful shapes. But other deciduous trees should be pruned every year for the first few years.

Inspect your young trees in late summer or late winter, and remove dead, damaged, and diseased wood. Weak or crossing branches are also best cut out. At the same time, loosen ties, and remove any

Water your tree thoroughly after planting, and then weekly during dry spells for the first two growing seasons – trees are most vulnerable to dehydration when young.

weeds growing over the root area. While you're at it, check for any suckers shooting from below the graft union and cut them out too (*see right*).

More extensive pruning depends on what kind of tree you want. Most trees sold at garden centres are lollipop-shaped, technically known as "standards". These have been pruned by the nursery, and you merely need to continue keeping the trunk clear, and pruning back wayward stems.

Younger trees, known as "whips", are also available. These usually have a single leading stem, and buds that will develop into side stems. They are less expensive, and often easier to establish than older trees, and they can be pruned and shaped as you wish. For example, if you want a tree with several stems, cut the stem to near ground level, or plant a few whips close together, or even in the same hole.

Pollarding restricts the size of a tree (here a catalpa) and encourages larger leaves. When the trunk of a newly planted tree reaches 1–2m (3–6ft), prune back the branches to 2.5–5cm (1–2in) in late winter or early spring. Then, every year, prune back the young stems that have regrown, and thin out any congested stems.

Coppicing is cutting back stems down to the base to stimulate new shoots; do this in late winter or early spring, just before the leaf buds break. Hazel and willow respond well to coppicing every year, but on poor soil it takes less out of the tree if you cut back half the stems one year, and the other half the following year.

REMOVING WATER SHOOTS AND SUCKERS

Remove water shoots and suckers as soon as you can, as they divert water and nutrients from the tree if left to grow. Suckers that grow from roots must be pulled off as close to the base as possible – pull back the soil to expose the area where it joins the root. If you cannot pull them off, use secateurs to remove them.

Water shoots that are growing through the bark or around the edges of a wound should be removed at their base; use secateurs to avoid tears in the bark.

Suckers need to be cut off as close to the base of the shoot as possible. Rub off any regrowth using your fingers or foot as soon as it appears.

PRUNING IN WINTER

It is easier to assess and prune deciduous trees in winter, when they are leafless. Exceptions are trees that bleed sap or are slow to heal, like birches and magnolias, and those prone to disease, such as prunus. Winter pruning results in lots of new shoots, so be ready to rub out unwanted growth.

Assessing your tree and making a pleasing shape is very rewarding work, but remember that although it is easy to remove a stem, it takes a good deal longer for it to regrow if you make a mistake, so take your time with the task and cut with care.

1 When the leaves of a deciduous tree have fallen, take a look at its overall shape first. Look for stems that are badly placed, or those growing too far down the trunk. This tree has an awkward stem growing from the base that must be removed. First, remove any dead and damaged wood. Then use a pruning saw to make a straight cut through any branches growing from the base of the tree.

2 Prune thin stems with loppers or secateurs, taking them back to 2cm (½in) from the ring of slight swelling where the stem and trunk meet, known as the "collar".

3 Thick branches, and those that are likely to tear, are cut in stages. First, cut under the stem, a short distance from the trunk. Cut about a quarter of the way through the underside of the branch.

4 Make a second cut above the lower one, and aim to join the two. Ensure your tools are sharp to prevent snagging.

5 Even if you have taken great care, a heavy branch may snap off and snag as it goes, but it is not so crucial at this point, since this is not the final cut.

6 Remove the stub by using the technique outlined in steps 3 and 4. Make your final cut just slightly away from the branch collar.

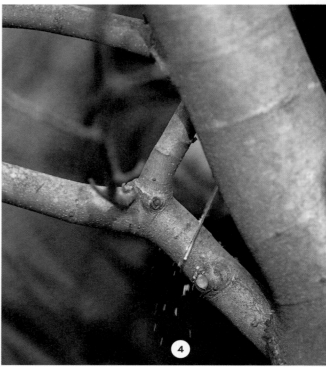

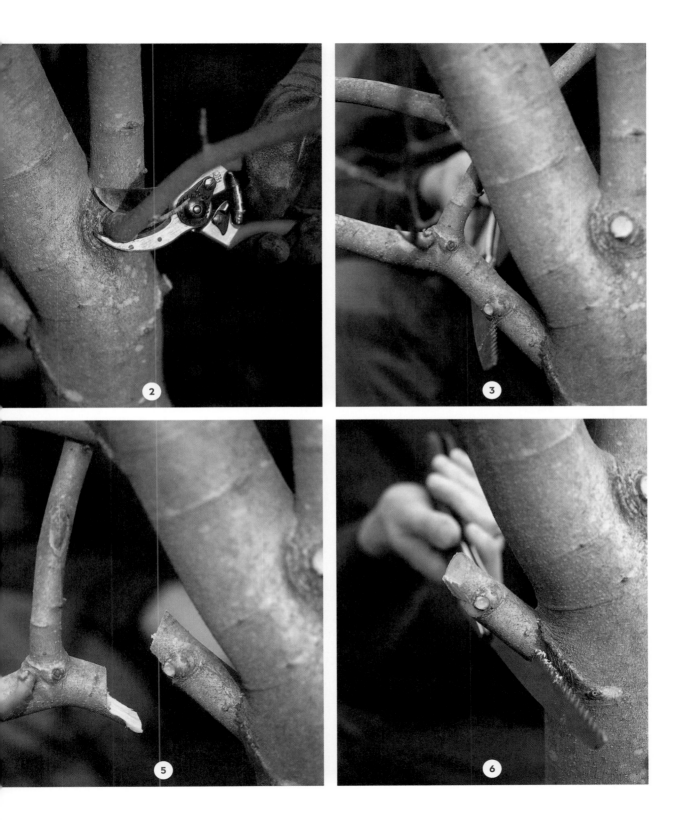

PRUNING IN SUMMER

Trees that bleed sap when cut, such as the magnolia shown in the pictures here, are best pruned in summer, when they will be more susceptible to disease but will recover more quickly – even though regrowth may be delayed while the tree concentrates its energy on healing its wounds. When pruning young magnolias for shape, you can lightly trim them in the spring, but don't cut stems any thicker than 2cm (½in) at this time.

1 Dead and diseased wood is quite obvious, so remove this first. Although it is more difficult to see the outline of a tree in leaf, it should be easier once you have removed any unhealthy stems.

2 Take out stems that are growing towards the centre of the tree. Be wary of pruning large branches that will heal slowly.

3 Remove any crossing branches to prevent them rubbing against each other and causing wounds that may result in serious damage, especially in the case of trees that bleed sap.

4 Prune out weak stems that did not produce flowers or that have few leaves, and rub out shoots forming on the lower trunk.

5 Keep the tree in shape by reducing the length of wayward side stems, cutting them back by about one third.

6 The pruned tree has a better shape and a more open canopy, allowing more light into the centre, and promoting vigorous stems.

TREES FOR SMALL GARDENS

1 *Acer palmatum* 'Nicholsonii' **Japanese maple**
Like all acers, grown largely for its delicate leaves and autumn colour.
H 2–3m (6–10ft); S 2–3m (6–10ft).

2 *Aralia elata* 'Variegata' **Japanese angelica tree**
Large, flat branches of green leaves with white margins. White flowers followed by black fruits from late summer.
H 10m (30ft); S 10m (30ft).

3 *Cercis canadensis* 'Forest Pansy' **Eastern redbud**
Spreading, deciduous tree with heart-shaped, dark purple leaves (yellow in autumn).
H 5m (15ft); S 5m (15ft).

4 *Sorbus aria* 'Lutescens' **Whitebeam**
A mid-sized, deciduous tree with silvery-grey young leaves and brown-speckled red berries.
H 10m (30ft); S 8m (25ft).

5 *Eriobotrya japonica* **Loquat**
An evergreen tree with long, heavily veined, glossy, green leaves and scented, white flowers in autumn, sometimes followed in spring by edible orange fruits.
H 8m (25ft); S 8m (25ft).

6 *Amelanchier lamarckii* **Juneberry** or **Snowy mespilus**
A small, bushy tree covered in clusters of white flowers in spring. Young leaves are silvery bronze, then green, and finally red in autumn.
H 10m (30ft); S 12m (40ft).

7 *Ficus carica* 'Brown Turkey' **Common fig**
Large, deciduous, green leaves and edible, purple fruits. An ideal tree for growing against a warm, sunny wall.
H 3m (10ft); S 4m (12ft).

8 *Arbutus unedo* **Strawberry tree**
Slow-growing evergreen tree with small, white flowers and strawberry-like fruits in autumn.
H 8m (25ft); S 8m (25ft).

TREES FOR FOCAL POINTS

1 *Pyrus salicifolia* 'Pendula' **Weeping silver pear**
Narrow, silvery leaves and profuse white blossom
are followed by small (but inedible) fruits.
H 5m (15ft); S 4m (12ft).

2 *Magnolia denudata* **Lily tree**
Spectacularly large, pure white flowers appear on bare
branches in spring, followed by large, glossy leaves.
H 10m (30ft); S 10m (30ft).

3 *Robinia pseudoacacia* 'Frisia' **False acacia** or
Black locust
This golden-leaved variety is smaller than the
standard species. White flowers.
H 25m (80ft); S 15m (50ft).

4 *Cornus alternifolia* 'Argentea' **Pagoda dogwood**
or **Green osier**
Small, deciduous tree or shrub that produces
spreading, horizontal tiers of green leaves with
white margins.
H 3m (10ft); S 2.5m (8ft).

5 *Betula utilis* subsp. *jacquemontii* **Himalayan birch**
A deciduous tree, elegant in all seasons but usually
grown for its paper-thin, white bark.
H 15m (50ft); S 8m (25ft).

6 *Stewartia monadelpha* **Stewartia**
Dark green leaves turn vibrant orange and red in
autumn before they fall. Bark is peeling red-brown
and grey.
H 25m (80ft); S 8m (25ft).

7 *Koelreuteria paniculata* **Golden rain tree** or
Pride of India
In summer, masses of yellow flowers appear, followed
by lantern-like, pink- or red-brown fruits. Leaves turn
yellow in autumn.
H 10m (30ft); S 10m (30ft).

8 *Taxus baccata* 'Fastigiata' **Irish yew**
Unusually narrow variety with erect branches and
dark green foliage.
H 10m (30ft); S 6m (20ft).

TREES FOR SPRING BLOSSOM

1 *Cercis siliquastrum* **Judas tree**
In spring a mass of magenta-pink, pea-like flowers appear all over the bare branches.
H 10m (30ft); S 10m (30ft).

2 *Malus* 'Red Devil' **Dessert apple**
Medium-sized apple tree yielding striking red fruits with pink-stained flesh in early autumn.
H 1.5–6m (5–20ft), depending on rootstock.

3 *Magnolia* x *soulangeana* 'Lennei Alba' **Magnolia**
Flower buds on the tips of branches open in spring to produce dramatic, pure white flowers 10cm (4in) across. H 6m (20ft); S 6m (20ft).

4 *Prunus* 'Spire' **Ornamental cherry**
Clusters of pale pink flowers appear at the same time as the leaves, in mid-spring.
H 10m (30ft); S 6m (20ft).

5 *Laburnum* x *watereri* 'Vossii' **Laburnum**
Spectacular showers of long, hanging, golden-yellow flower clusters in late spring.
H 6m (20ft); S 6m (20ft).

6 *Cornus kousa* 'Miss Satomi' **Japanese dogwood**
Deciduous tree or shrub with masses of dark pink flower bracts in late spring or early summer.
H 7m (22ft); S 5m (15ft).

7 *Crataegus laevigata* 'Paul's Scarlet' **Hawthorn**
A thorny, deciduous tree grown for its deep pink double flowers in late spring.
H to 8m (25ft); S to 8m (25ft).

8 *Malus* x *arnoldiana* **Crab apple**
In mid- to late spring, red buds open to scented flowers, initially pink, fading to white.
H 5m (15ft); S 8m (25ft).

TREES FOR
AUTUMN COLOUR

1 *Acer palmatum* 'Sango-kaku' **Coral-bark Japanese maple**
Leaves are yellow in spring, then green, with a blaze of golden orange in autumn.
H 6m (20ft); S 5m (15ft).

2 *Ginkgo biloba* **Maidenhair tree**
Unusual, fan-shaped, green leaves turn an attractive golden-yellow in autumn.
H to 30m (100ft); S to 8m (25ft).

3 *Liquidambar orientalis* **Oriental sweet gum**
A small variety whose maple-like leaves turn fiery crimson, yellow and orange in autumn.
H 6m (20ft); S 4m (12ft).

4 *Sorbus commixta* **Japanese mountain ash** or **Japanese rowan**
A neat, compact sorbus. White flower clusters in late spring, then red berries and good autumn colour.
H 10m (30ft); S 7m (22ft).

5 *Sorbus sargentiana* **Sargent rowan**
White flowers give way to red berries, and bright orange and crimson autumn foliage.
H 10m (30ft); S 10m (30ft).

6 *Nyssa sinensis* **Chinese tupelo**
Slender leaves are bronze when young, dark green in summer, then many-hued in autumn.
H 10m (30ft); S 10m (30ft).

7 *Parrotia persica* **Persian ironwood**
A spreading, deciduous tree with dense foliage turning yellow and red-purple in autumn.
H 8m (25ft); S 10m (30ft).

8 *Rhus typhina* **Stag's horn sumach** or **Velvet sumach**
Stems are velvet-coated (like antlers). In autumn leaves are rich ruby, gold, and amber.
H 5m (15ft); S 6m (20ft).

HEDGES

Hedges have many uses in the garden. Most commonly planted as simple boundary screens, they can also serve as internal divisions or low edging for beds, borders, or paths. When close-clipped and uniform, hedges create a sense of smart formality; when mixed and billowy, they are a garden in themselves. Hedges have other practical uses, too. They make excellent windbreaks – more effective, in fact, than solid walls and fences. Deciduous hedges, in particular, filter the blast and slow windspeeds, whereas solid barriers actually cause turbulence on the sheltered side. Evergreens help muffle the noise of traffic and offer year-round privacy, while dense, thorny hedges deter intruders.

Trimming hedges is essential if you want to maintain a well-kept, formal look. It will probably need doing at least twice a year – once in midsummer and again in autumn.

GARDEN HEDGES

All hedges take some years to establish and gain the required height but some, such as oval-leaved privet (*Ligustrum ovalifolium*), are relatively fast-growing. Remember, though, that the faster they grow, the more frequently they will need clipping. While your hedge is maturing, you can erect a temporary screen of split bamboo or hurdles, or a wire-netting fence as a boundary line. This netting can be left in place as it will eventually be covered by the hedge. If you can't wait, "instant hedges" (grown in containers) can be bought by the metre. They are not cheap but the contracting company will also prepare the ground and carry out the planting.

Use hedges as garden dividers where you want to split up your space into different rooms. Tall hedges here create a secret, sheltered enclosure, further "furnished" with dwarf box edging.

WHAT SORT OF HEDGE?

Evergreens give cover all the year round while deciduous hedges are see-through in winter, except for beech and hornbeam, which retain crisp brown leaves until spring. If you live in a cold, exposed place, choose a hardy, deciduous hedge that will withstand windy conditions, such as hawthorn (*Crataegus monogyna*) or hornbeam (*Carpinus betulus*). Plants for seaside gardens exposed to wind, but relatively frost-free, include the hawthorn and hornbeam again, and also the sea buckthorn (*Hippophae rhamnoides*), and tamarisk, and the evergreens griselinia, holly, and olearia, all of which tolerate salty air.

For a smooth, formal evergreen hedge, look for small-leaved plants with a dense growth habit that tolerate regular close clipping. Two of the best choices for formal hedges are yew (*Taxus baccata*) and Lawson cypress (*Chamaecyparis lawsoniana*). Others to consider include box (*Buxus sempervirens*), or fast-growing privet (*Ligustrum ovalifolium*), and shrubby honeysuckle (*Lonicera nitida*).

Informal hedges are often a mixture of flowering and berrying, evergreen, and deciduous shrubs planted closely together. They are well suited to rural or semi-rural areas and informal garden settings, and need little or no pruning if they are to show off their beauty, so allow them a bit more space than a clipped hedge. You can achieve a more formal look by mixing plants of a similar growth rate and size, but with different foliage colours, such as green and copper beech, or green and golden yew.

EXISTING HEDGES

If you already have a hedge, you may consider it your pride and joy, or merely a chore. If your current hedge is not doing anything for you or your garden, you could dig it up and start again. This is not as drastic a step as it may seem – yes, hedges are long-lasting, but they are also renewable and some grow faster than you might think. You could replace your old hedge with a low-maintenance version – beech (*Fagus sylvatica*), for example, needs just one trim per year – or, if space is limited, you could achieve a similar effect by training ivy through wire netting or trellis to make a narrow divide.

(top) **Dwarf box and beech** hedging have been combined here to create an exciting textural contrast, while the spiky fern infill offers yet more visual and tactile interest to this framework of foliage.

(bottom) **Closely clipped hedges** provide beautiful natural backdrops to herbaceous borders, forming an unbeatable contrast for billowing, colourful planting such as this.

PLANTING LEYLAND CYPRESS

Think twice before you plant Leyland cypress (x *Cupressocyparis leylandii*), as it has gained a bad reputation in the gardening world. Left untrimmed, it quickly grows to a great height. If it is then cut back hard it looks unsightly – brown stems will stay twiggy, and will not regain their former green colour. Ideally, these hedges should be lightly clipped to shape about twice a year.

This vigorous grower also dries out the surrounding soil and depletes it of nutrients, and should be discounted as a boundary for small gardens, or it will tower over all the other plants, and block out the sunlight. It is most useful as a windbreak in large gardens or in coastal areas.

HEDGES FOR WILDLIFE

Hedges can enrich the garden by encouraging birds and other wildlife, providing shelter, nesting sites, and food in the form of berries and the insects they harbour. A wildlife hedge of mixed species, such as field maple (*Acer campestre*), hawthorn (*Crataegus monogyna*), holly (*Ilex aquifolium*), hazel (*Corylus avellana*), and spindle tree (*Euonymus europaeus*) will attract visitors all year. But don't clip it too hard if you want flowers and fruit. If space allows the occasional plant to grow into a small tree, so much the better. For extra flowers, plant species roses, which also offer hips in autumn, and viburnums.

CONSIDERATE PLANTING

Hedges can cause serious conflict between neighbours. Firstly, bear in mind that they take up much more space than walls or fences, and are unsuitable for the boundaries between very narrow gardens.

To stay on good terms with your neighbours, keep them informed of your boundary plans and do not allow your hedge to encroach on their garden. At planting time, ensure there is sufficient space for the hedge to spread, and, by mutual agreement, be prepared to cut it on both sides. Also, ensure that your hedge does not grow too high and shade your neighbour's property. Boundary hedges can be allowed to grow up to 1.8m (6ft). Internal hedges for screening parts of the garden can be taller, but think about how you will clip them before they get out of hand.

Root-balled hedging plants are more expensive than bare-root plants, but cheaper than those offered in containers.

Bare-root transplants should be carefully separated and planted out immediately, to prevent the exposed roots from drying out.

BUYING HEDGING PLANTS

Success with a new hedge depends on preparing the soil thoroughly before your plants arrive, and pampering them after planting and in the early years. Choose young plants rather than large specimens, since they will grow faster and soon overtake mature plants, giving you a much better hedge for a fraction of the cost.

Nurseries specializing in hedging offer a wide choice of plants, and many do mail order. A good supplier will tell you how plants are packed for transit, and offer planting and pruning advice. Find suppliers on the internet or through adverts in gardening magazines.

Plants supplied by specialists are offered in three forms: bare-rooted transplants, root-balled, or in containers. Bare-rooted transplants are much cheaper than container-grown specimens; mixed native hedging is usually the most cost-effective option. Bare-root transplants are available from autumn to spring, but you can order them at any time of the year and request a delivery date. Remember to ask for delivery in spring if you are buying evergreens

or conifers and live in a cold, windy area, since this is the best planting time. Evergreens are also available root-balled, where the plant is lifted from the field with its roots and some soil, and then wrapped to reduce root disturbance.

Always buy more plants than you need in case some of your hedging dies. Plant the surplus in a quiet area of the garden or in containers that are sufficiently large to allow the roots to grow. Keep these plants well watered until your hedge is established and you are sure none need to be replaced.

Bare-rooted and root-balled hedging plants can be bought from autumn to spring, but will not usually be lifted from the field in very severe weather, so let availability be your guide. In cold areas, conifer and evergreen hedges in particular will fare better if planted in spring, rather than winter. Pot-grown plants can be bought and planted all year, but it makes sense to plant in mild, moist autumn or spring conditions. If you can't plant as soon as you receive your hedging, keep plants moist in their packing, and store in a cool shed.

PLANTING A HEDGE

A single row of plants is sufficient for most hedging purposes, but for an extra-thick hedge a double row can be used, provided that you have enough space. In this example, a single row is being planted hard up against a temporary hurdle fence. Spacing along the row for most hedging plants, whether they are deciduous or evergreen, should be between 45–60cm (18–24in), but for smaller-leaved privet, box, and *Lonicera nitida*, plant closer, at 30cm (12in). For dwarf edging hedges of box or lavender, for example, plant 15cm (6in) apart.

1 Start preparing the ground at least six weeks before planting. Clear the site of perennial weeds, then dig a strip the length of the hedge and 90–120cm (3–4ft) wide. Fork plenty of bulky organic matter (such as garden compost or well-rotted manure) into the trench as you go.

2 Set out a line to mark the edge of the row(s) of plants. Dig a trench along it for bare-root plants; you can dig individual holes for containerized or root-balled plants after marking out their positions. Use canes and a ruler to mark out the planting intervals, either in a single row along one line, or in two staggered rows, laying canes 45–60cm (18–24in) apart as a guide.

3 Plant at the original planting depth – this will either be the top of the rootball, or to the soil mark on the stem of bare-root plants.

4 Return the soil to the trench, and firm it well using your heel, at the same time creating a slight dip around the plants.

5 Water the plants thoroughly using the dip (see *pp.198-199*) around each plant to help direct water to the roots where it is really needed.

6 Place a thick mulch of well-rotted manure or compost, or of shredded bark, around the plants, making sure it stays well clear of the stems. Alternatively lay landscape fabric in strips on each side and cover with chipped bark.

CARING FOR HEDGES

Once planted, your hedge will need looking after if it is to thrive. Most importantly, keep young plants watered for the first few years during dry spells until the hedge has become established. It should then not need additional watering except in dry periods in the summer. Feed all hedges in spring with a balanced fertilizer, such as blood, fish, and bone. At the same time, regularly clear any weeds from the base of the hedge, and after rain, or when the soil is moist, top up any mulch.

CLIPPING AND SHAPING

How you shape your hedges will depend on their age and whether you have grown them to be formal or informal.

• **Young informal hedges** need no shaping, although shortening any long, whippy shoots will encourage them to bush out.

• **Young formal hedges** may be pruned to shape and direct growth. Evergreens need only their sideshoots lightly trimmed for a few years after planting. Leave the main upright, or leading, shoot(s) to grow to the final hedge height. Deciduous hedges that are to have a dense, formal shape require more early shaping. In general, cut back the main shoot and any strong sideshoots by one-third after planting; repeat the following winter. Prune vigorous privet and hawthorn harder: cut back to within 30cm (12in) of soil level in late spring. In late summer shorten sideshoots; then, in winter, cut the previous season's growth back by half.

• **Mature informal hedges** should be merely trimmed in spring or summer, and left unpruned unless they are outgrowing their space or looking particularly untidy. Snip off any over-long shoots and keep the height under control.

• **Mature formal hedges** need keeping in shape with regular clipping. This keeps growth dense, and the hedge thick and lush from top to toe. It also prevents it from expanding over the years. How often you clip depends on the plants and their rate of growth, but all hedges need at least one cut a year. For deciduous hedges and small-leaved evergreens, use garden shears or a powered hedge trimmer, keeping the blades parallel to the hedge at all times for flat sides. For large-leaved evergreens, such as cherry laurel (*Prunus laurocerasus*) and tough-leaved holly, use secateurs to cut back individual shoots.

(left) **Make a trim line** by tying string to two canes; insert them at each end of your hedge to make a clipping guide for formal hedges, such as box.

(right) **Hedges with large, thick leaves,** such as holly and laurel, should be trimmed with secateurs; shears will cut the foliage, turning it brown.

(top) **When using a powered hedge trimmer,** wear goggles, ear defenders and stout pruning gloves. Use a model you can handle with ease.

(bottom) **Fill gaps** in a hedge base by layering a stem. Choose a low stem, and allow it to grow until it easily reaches the ground and can be pinned under the soil to root.

FILLING IN GAPS IN A HEDGE

If a plant in a young hedge dies, dig it out and plant a new one. Switch some of the soil with some from another part of the garden as a precaution, enriching it with well-rotted compost or manure. In an established hedge, remove the dead plant, and fill the gap with shoots of adjacent plants trained horizontally along strings or wires set between canes.

If a stem is forced out of position by snow or wind, tie it back in place using soft twine. Clipping your hedge into an A-shape (see *right*) will reduce the risk of this happening. Cut out browning foliage, particularly on conifers, as it will not green up. If it leaves a gap, tie adjacent branches together with soft twine. Keep an eye out for more dead patches, which may indicate disease.

RENOVATING AN OVERGROWN HEDGE

Where an established hedge is overgrown, start by cutting back weeds such as ivy and brambles. When new growth sprouts, paint the leaves with glyphosate weedkiller – more than once if needed. Remove any dead wood and cut the hedge back. Apart from when dealing with conifers, spread this over two years – cut back the top and one side one year, and the other side the next.

Beech, hornbeam, and hawthorn can be cut back hard in late winter before new leaves appear. If regrowth is poor in the first season, delay the second half of the job for another year. For an old forsythia hedge, cut out some of the old wood once flowering is over. Among evergreens, yew, box, cherry laurel, privet, lonicera, holly, escallonia, evergreen berberis, and pyracantha can cope with hard pruning, in mid- to late spring when growth is well under way. After pruning, feed and water, and mulch with bulky organic matter.

CLIPPING HEDGES INTO SHAPE

Prune formal hedges, such as beech and yew, from the time of planting so that the plants form a wedge or A-shape as they grow. These shapes allow light to reach all areas of the hedge, which will encourage even growth.

If you live in an area that experiences snow, clip the top into a point or narrow A-shape so that the snow cannot settle on the top where its weight may damage the top of the hedge.

For a hedge with a narrow profile, clip the sides regularly from the start. Trim it back each time to the last cut, for a smooth finish.

Beech with a pointed top to shed snow.

"A-shaped" yew hedge with a flattened top.

PLANTS FOR HEDGES

1 *Carpinus betulus* **Hornbeam**
A deciduous tree that does better than beech in exposed sites, and holds its juvenile dead leaves in winter. Trim in mid- to late summer.
H to 9m (30ft).

2 *Lonicera nitida* 'Baggesen's Gold'
Shrubby honeysuckle
A "Dalek"-shaped evergreen shrub, best grown as a formal hedge. Fast growing and may need several trims a year.
H to 1.5m (5ft).

3 *Prunus lusitanica* **Portugal laurel**
A dense, evergreen laurel with glossy, dark green leaves. For a semi-formal hedge, trim with secateurs in late summer.
H to 20m (70ft).

4 *Ilex aquifolium* 'Golden Milkboy' **Holly**
Slow-growing, dense evergreen, clipped with secateurs in late summer for formal hedging. This variety has irregular yellow patches on the leaves.
H 6m (20ft).

5 *Lavandula angustifolia* 'Hidcote' **Lavender**
Best as low, semi-formal edging. Trim with shears in spring and after flowering to keep compact.
H to 60cm (24in).

6 *Berberis darwinii* 'Barberry' **Barberry**
Vigorous evergreen covered in a mass of dark orange flowers in spring. Trim after flowering to encourage more flowers in autumn.
H to 3m (10ft).

7 *Taxus baccata* **Yew**
A formal and slow-growing evergreen that may need only one clip a year in summer or autumn.
H to 10m (30ft).

8 *Ligustrum ovalifolium* **Oval-leaved privet**
Fast-growing shrub best trimmed into formal hedging three or four times in the growing season. Normally evergreen, though it may lose leaves in cold winters.
H to 4m (12ft)..

9 *Chamaecyparis lawsoniana* 'Pembury Blue' **Lawson cypress**
Quite fast growing cypress tree with sprays of blue-grey foliage. Trim in spring and early autumn.
H to 9m (30ft).

10 *Fagus sylvatica* **Beech**
A deciduous tree that keeps its juvenile brown, dead leaves in winter. Mix green- and copper- or purple-leaved varieties for a "tapestry" effect. Trim in late summer.
H to 9m (30ft).

11 *Buxus sempervirens* 'Latifolia Maculata' **Box**
A dense, small-leaved evergreen that needs trimming two or three times in the growing season for formal hedging. This is a variegated form.
H to 5m (15ft).

12 *Rosa rugosa* 'Rubra' **Hedgehog rose**
Grows vigorously and densely, and is the best rose for a tall, informal hedge. Very prickly stems, and large red hips in autumn.
H 1–2.5m (3–8ft).

CONTAINERS

Plants growing in containers add colour and
interest to your garden, as well as broadening the range
of plants you can grow. Containers inject drama into
areas where planting in open ground is not possible.
Well-chosen plants enhance paved areas, softening
hard brick and concrete surfaces, and bright flowers
or foliage can bring colour to dark corners. Suitably
planted containers can reinforce the style of a garden, too.
An olive tree, yucca, or palm in an urn may create a
Mediterranean feel, while colourful plants in vivid
containers are perfect for creating a vibrant look.

Mixed plantings usually look best if you avoid a kaleidoscope
of wildly different colours. Here, a container of *Cosmos bipinnatus*
'Sonata Pink', *Gypsophila muralis* 'Gypsy Pink', and a variegated
ivy employs a simple colour scheme of pinks and greens.

GROWING PLANTS IN CONTAINERS

Containers allow you to grow plants that dislike your garden conditions. For example, if you have alkaline soil but would like shrubs that need acidic soil, such as azaleas and pieris, they will thrive in pots of ericaceous compost. If your soil is dry, you can keep astilbes and hostas moist. Tender plants, such as agaves, which prefer dry, well-drained sites and are not reliably hardy in most areas, can be grown in pots even in gardens with wet or waterlogged soil, and moved under cover in winter. And plants with different soil needs can be grouped together in individual pots to create combinations that would be impossible in the open ground.

WHERE TO PLACE CONTAINERS

Container plants perform best in a sheltered spot, so patios and terraces are ideal locations. In exposed sites, pots dry out quickly and may blow over. Just as in the open ground, it is important to put container plants in the right place, grouping those that like similar conditions to achieve healthy displays.

Use plants with strong shapes, such as cordylines, to create an instant focal point. Grown in containers that can easily be moved, they offer a quick way to change emphasis.

Displays are simple to create and change. Large pots can be used as focal points; pairs of identical pots with matched plantings are ideal either side of paths, and if placed to each side of a doorway can emphasize architectural features. Conversely, containers can hide eyesores, such as manhole covers or drains.

Fill seasonal gaps in the border with potted lilies or tulips, which bloom spectacularly, but only for a limited period. Once their flowers fade, the pots can be lifted and replaced. A group of containers holding ferns, hostas, and other shade-tolerant plants should survive underneath established trees, where the ground is typically dry, poor, filled with roots, and difficult to cultivate.

If your outdoor space is paved, containers are ideal. Some climbers, such as jasmine and certain clematis, grow well in pots, so even on a concreted area can be trained to scramble up walls. Even with no garden, you can grow plants in containers. Site windowboxes on wide window ledges, or hang them from the wall on brackets. Hanging baskets can be placed in a sheltered spot where they are accessible for maintenance and watering. Wall-mounted pots can brighten a bare wall, although they only hold a small amount of compost, so require copious watering in summer. For all these containers, robust fixings are vital.

CHOOSING CONTAINERS

1 Terracotta and clay are versatile materials, with pots available in a range of shapes and styles: urns, vases, bell pots, pans, bowls, and jars. Plants grown in jars are tricky to repot if the rootball is wider than the pot's neck. Planting holes in the sides are often used for herbs and strawberry plants.
Pros: Porous terracotta allows the rootball to breathe, reducing waterlogging. Huge range of colours and styles available, often relatively inexpensive.
Cons: Porous nature of terracotta means plants need frequent watering. Often broken when tipped over or due to frost shattering.

2 Synthetic containers such as plastic, fibreglass, and resin are widely available alternatives to traditional materials. They are lightweight, frost-proof, and almost unbreakable. Large synthetic pots are easily moved, so are ideal for tender plants, which must be in a sheltered place in winter, and decorative foliage plants, which can be moved to mask flowers that are past their best.
Pros: Frost-proof, lightweight, and inexpensive, and available in a range of colours and styles.
Cons: Poor insulation, and easily blown over.

3 Metal containers come in a range of shapes, colours, and styles; many are suited to modern settings. May be galvanized, painted, or coated to protect against rust, though some finishes gradually tarnish. A liner will insulate roots against frost.
Pros: Wide-range of styles, robust, and attractive.
Cons: Good-quality pieces are expensive. Lead containers are heavy. Shiny finishes show up mud splashes, and will dull.

4 Natural wood planters are unobtrusive and provide a perfect foil for plants. Wood is light and easy to handle, but must be treated if it is to last. Half barrels make excellent large containers.
Pros: Not easily broken, and durable if treated. Also lightweight with good insulating properties.
Cons: May rot if not treated or emptied regularly.

5 Stone containers are among the most desirable and durable types of container. Stone urns suit a traditional setting, especially once aged, making them ideal focal points in a period garden. Contemporary designs are also available. Reconstituted stone is cheaper and more widely available.
Pros: Durable, weighty, and stable, with good insulating properties. Available in many styles.
Cons: Both heavy and expensive.

PLANTING UP CONTAINERS

To ensure that plants grow and perform as well as possible, you need to plant them properly. Your choice of compost will depend on what you are planting:

• **Soil-based compost** is better for long-term plantings like specimen shrubs; it holds nutrients for longer, and drains better – important if plants are to survive outside in winter.
• **Soilless compost** is usually best for short-term planting, such as bedding plants for summer containers. It is light and clean, and has nutrients for young plants; you may have to add fertilizer later. Also ideal for hanging baskets.

If your chosen pot is large or heavy, place it in position first, and then plant it. Check that the drainage holes are sufficient, then fill the bottom of the pot with extra drainage material, such as polystyrene pieces or crocks, and cover with a 2.5–5cm (1–2in) layer of compost.

1 Water the plant thoroughly about an hour before planting, then position it – still in its original pot – inside the new container.

2 Add compost to raise the plant to 2.5cm (1in) below the rim; this allows space for watering. Also add compost around the sides.

3 Firm the compost well around the pot. Lift out the plant in its pot, leaving a hole in the compost. Remove the plant from its original pot.

4 If there is a tight mass of roots around the rootball, tease them out gently. Place the plant back into the hole in the compost, and firm in.

5 Check that the plant is at the same depth as it was in its old pot. Then cover the top with a little more compost, and water well.

6 Finish off with a layer of ornamental mulch. This helps to retain soil moisture and suppresses weeds, as well as anchoring the pot.

CARE AND MAINTENANCE

Plants in containers are easy to maintain and control. Feeding and watering can be regulated and monitored, while deadheading and pruning are simple, as plants are moveable. When a plant's season of interest has passed, you can simply replace it. Plants that grow too vigorously in open ground, such as some bamboos, are restrained by pots, while slower-growing species that are easily swamped in a border can be pampered. Pests and diseases are easy to treat, and plants can be lifted above soil level away from slugs and snails. However, plants in pots depend on you totally for their food, water, and general care, whereas the same plants in open ground will survive with far less attention.

WATERING

Containers dry out quickly, especially in hot or windy weather. Some plants, like camellias and rhododendrons, do not set flower buds properly if they go short of moisture. Being evergreen, they need watering all year. Pots close to walls and fences or under trees do not receive much rainfall, so keep an eye on them. The signs of lack of water are wilting, yellowing foliage, and in extreme cases, falling leaves.

Use a watering can or a hose fitted with a lance or gun so you can moderate the flow of water. The plant has had a thorough soaking once water flows from the bottom of the pot. Surface compaction may prevent water soaking in at all; push a hand fork into the soil a few times to remedy this. An irrigation system is the most reliable way to keep plants watered. Small bore pipes with adjustable nozzles are placed in the pots; water can be controlled by a timer.

(far left) **A piece of slate** or crock can be used if you don't have a watering lance. Place it on the soil as you water, and it will help to distribute the flow evenly.

(left) **When a plant has dried out,** compost shrinks away from the sides of the pot, and water runs through the gap. To alleviate this, submerge the pot in water for up to an hour.

(far left) **Water-retaining gel** is useful for those who occasionally forget to water; it is excellent for hanging baskets, which dry out quickly. The gel stores moisture around the rootball, creating a reservoir for the plant to draw on in dry periods. Add the gel granules to the compost at planting time, following the directions on the packet.

(left) **Adding water-soluble fertilizer** to a watering can is the easiest way to feed your plants. Always follow the instructions on the packet, as an excess of fertilizer can kill plants.

FEEDING

Plants in containers need feeding from late spring to mid-autumn to produce healthy growth and a succession of flowers. Feeding is usually done every fortnight, although hungry plants, such as tomatoes and brugmansias, need food weekly.

The simplest method is to add water-soluble feed to the watering can. Controlled-release fertilizers avoid the need for continual feeding, as the fertilizer is released over several months. These are available as pellets, sticks, or small plugs, which are pressed into compost once the plant is in the pot. Alternatively, buy granules and mix them into the compost before planting. All forms are useful in a low-maintenance garden. Controlled-release fertilizers are also the best option for containers that are packed full of plants, such as hanging baskets. Always follow the manufacturer's guidelines; overfeeding can be fatal to plants.

Adding well-rotted manure to the bottom of the pot when planting or sprinkling pelleted chicken manure on pots in spring are good organic alternatives.

Moving heavy pots, even when empty, is not easy. Grow plants that will be moved regularly in light pots and soilless compost. Plant movers – metal or wooden stands on wheels, and fitted with brakes – are useful for heavy containers.

REPOTTING

Repot regularly in spring, using a new pot only slightly larger than the original. Pot-bound plants are often top-heavy, and hard to water, with little space for water to collect on and seep into the soil without spilling over the edge or running down the inside of the pot. Foliage may be yellow, and new growth reduced. If you want to keep an overgrown plant in the same pot, either prune it, or divide it and replant in fresh compost (see *pp.304–305*). Also, top-dress potted trees and shrubs in spring by removing one third of the compost and replacing with fresh potting compost.

FLOWERS AND CLIMBERS FOR CONTAINERS

1 *Brachyscome iberidifolia* 'White Splendour'
Swan river daisy
Spreading annual, ideal for hanging baskets.
White, pink, or purplish flowers.
H to 45cm (18in); S 35cm (14in).

2 *Thunbergia alata* **Black-eyed Susan**
Half-hardy climber with bright orange or yellow
flowers. If grown as a perennial, protect pots in winter.
H to 2.5m (8ft).

3 *Tropaeolum peregrinum* **Canary creeper**
Vigorous annual climber bearing bright yellow
flowers with fringed petals from summer to autumn.
H 2.5–4m (8–12ft).

4 *Hyacinthus orientalis* 'Blue Jacket' **Hyacinth**
Plant bulbs in autumn for richly scented flowers
in mid-and late spring.
H 20cm (8in); S 30cm (12in).

5 *Dahlia* 'Yellow Hammer' **Dahlia**
A dwarf bedding dahlia suited to pots.
Single, yellow flowers and bronze leaves.
H 60cm (24in); S 45cm (18in).

6 *Pelargonium* 'Lemon Fancy' **Pelargonium**
Purple flowers and citrus-scented leaves.
Protect from frosts.
H 40–45cm (16–18in); S 15–20cm (6–8in).

7 *Plumbago auriculata* **Cape leadwort**
Evergreen climber. Grow in a sunny spot; bring
pots indoors in winter.
H 3–6m (10–20ft); S 1–3m (3–10ft).

8 *Cobaea scandens* **Cobaea**
A tender, perennial climber usually grown as an
annual unless protected in winter. Fragrant, white
flowers turn purple as they age.
H 15m (50ft).

TREES AND SHRUBS FOR CONTAINERS

1 *Choisya ternata* 'Lich'
Mexican orange blossom
An evergreen shrub with lovely foliage –
yellow when young, turning green with age.
H 2.5m (8ft); S 2.5m (8ft).

2 *Fuchsia* 'Thalia' **Fuchsia**
A compact variety with small, scarlet flowers that
have unusually long tubes. Bring indoors in winter.
H to 75cm (30in); S to 75cm (30in).

3 *Rhododendron* 'Irohayama' **Azalea**
A dwarf azalea with pale pink or lavender flowers.
Needs special acid (ericaceous) potting compost.
H 60cm (2ft); S 60cm (2ft).

4 *Fatsia japonica* **Japanese aralia**
A large, striking, architectural plant with
wide-spreading, glossy, green leaves. It needs
a correspondingly large container.
H 1.5–4m (5–12ft); S 1.5–4m (5–12ft).

5 *Trachycarpus fortunei* **Chusan palm**
A large, evergreen palm that can withstand
cool temperatures, but is safer protected
indoors in winter.
H 3m (10ft) or more; S 2.5m (8ft).

6 *Hydrangea macrophylla* 'Nikko Blue' **Hydrangea**
Mophead variety with rounded, blue blooms.
H to 1.5m (5ft); S to 2m (6ft).

7 *Camellia japonica* 'Ace of Hearts' **Camellia**
Evergreen, flowering in early spring. Needs a
large pot, acid compost, and a sheltered site.
H to 3m (10ft); S 2m (6ft).

8 *Salix caprea* 'Kilmarnock' **Kilmarnock willow**
A weeping willow small enough for a large
container. Spring catkins before the leaves.
H 1.5–2m (5–6ft); S 2m (6ft).

GROWING CROPS IN CONTAINERS

Pots and containers can turn a small plot, patio, or courtyard into a miniature kitchen garden. In fact, there are numerous advantages to growing fruit and vegetables this way. Containers can be moved, allowing you to exploit microclimates – they can be placed somewhere open and sunny in summer and brought back to a sheltered spot in cold or windy weather. You also have control over the quality of the soil or potting compost you use.

CHOOSING CONTAINERS

Almost any kind of container can be pressed into service for growing crops – from beautiful old terracotta pots to improvised planters made from salvaged wooden boxes or plastic crates. The single, most important rule is that there must be a hole or holes in the bottom, to allow drainage and prevent waterlogging.

Clay pots are authentically traditional and will weather beautifully. They are also less likely to tip over if plants such as climbing beans or tomatoes become top-heavy. But they can be fragile, and may crack or flake in heavy frosts. Plastic pots are less expensive. They are also lighter, so can be moved more easily from place to place around the garden, and they are impermeable, so the soil within will not dry out so quickly. However, the environmental cost of single-use plastic means that containers made from the material should be reused as much as possible, and recycled when they are no longer usable. Wooden containers tend to rot where they come into contact with wet soil, so either treat them with preservative before filling them or place plastic liners inside.

Raised beds can be thought of as containers, in a sense: though they can't be moved, they do give you some control over the quality and composition of the soil (see *pp. 244–45*).

A miniature herb garden featuring a mixture of sage, purple sage, lavender, and parsley puts a stack of old car tyres to practical use.

Strawberries can easily be grown in pots (and the fruit stays cleaner) but they must be kept well watered.

Hanging baskets suspended from olive tree branches make attractive aerial herb and salad gardens, as long as you keep them regularly watered.

GROWING VEGETABLES

The vegetables that do best in pots tend to be those with fairly shallow roots, such as fruiting vegetables (see *pp.258–59*) and, if they are supported, climbing French and runner beans (see *pp.256–57*). Salad crops are particularly suitable, especially lettuces, endives, corn salad, spring onions, radishes, and cut-and-come-again leaves. Even potatoes can be grown in deep containers. Many garden centres sell special dwarf or patio vegetable varieties that are ideal. The vegetables least likely to do well are perennials such as asparagus and brassicas such as broccoli or cauliflowers.

GROWING FRUIT AND HERBS

Growing fruit trees in pots is always going to be a challenge; you'll need very large containers and plenty of space. But don't rule it out. There are numerous varieties of apple, pear, cherry, and even peach and apricot available that have been bred specially for container-growing and will therefore not become too tall or spread too wide. Indeed, citrus fruits such as lemons, limes, and oranges were traditionally grown in large pots that were brought into glasshouses or "orangeries" in winter and moved outdoors in summer. Figs actually fruit better in pots, blueberries can be grown in the special acid (ericaceous) soil they require, and strawberries can be better protected from snails and slugs.

Herbs are perfect for container growing, too (see *pp.261–63*). Rosemary, basil, thyme, marjoram, oregano, and parsley all do well, if kept watered, and may be more convenient for use if grown in pots close to the kitchen door or windowsill than planted out in the open ground. Mint is particularly appropriate for container growing. Left to its own devices, it is very invasive and will spread uncontrollably; constraining it to a pot keeps it firmly in check.

PLANTING A HANGING BASKET

Summer baskets are usually planted in late spring, but most of the plants suited to baskets are tender, and need protection until frosts have passed. Decide what kind of liner to use, and ensure it is the correct size for the basket.

1 Sit the basket in a pot or bucket to keep it stable throughout the planting process. Put the liner (here made of coir) in position and trim it to fit along the rim. A circle of polythene placed at the bottom will help retain water.

2 Cut a series of slits for planting trailing plants through the sides. Add a 5cm (2in) layer of soilless compost, mixed with controlled-release fertilizer and water-retaining gel.

3 Remove the trailing plants from their pots and thread them through the slits you have just created. It's a good idea to reuse a plastic bag to protect the stems and leaves.

4 Place a pot in the centre at the top of the basket to act as a watering reservoir, and plant around it. Use upright plants in the centre and trailers around the rim. Add more compost and build up the planting in tiers until the compost is within 5cm (2in) of the rim.

5 Firm the compost, and water the basket thoroughly through the plastic pot. Use a watering lance so that the water doesn't puddle in one section.

6 Hang your basket in a sheltered spot at a height that you can reach easily to water it. Ensure that the fittings are strong enough to take the weight of saturated compost.

PLANTS FOR HANGING BASKETS

1 *Lobelia erinus* 'Colour Cascade' **Lobelia**
Annual trailing lobelia produces a summer-long profusion of small blue, white, pink, red, and violet flowers.
H 15cm (6in); S 10–15cm (4–6in).

2 *Hedera helix* 'Little Diamond' **Variegated ivy**
A miniature, trailing ivy with grey-green leaves that have creamy-white margins.
H 30cm (12in).

3 *Diascia barberae* 'Blackthorn Apricot' **Diascia**
Apricot-coloured flowers are produced through summer and well into autumn.
H 25cm (10in); S to 50cm (20in).

4 *Glandularia* 'Diamond Merci' **Verbena**
Rounded clusters of small, burgundy-red flowers from midsummer to autumn.
H 30cm (12in); S 30–50cm (12–20in).

5 *Felicia amelloides* 'Read's White' **Felicia**
A white variety of the more usual blue daisy. It should flower all summer.
H 60cm (24in); S 60cm (24in).

6 *Lotus berthelotii* **Parrot's beak** or **Coral gem**
Trailing, evergreen shrub with pairs of eye-catching, orange-red flowers like lobster claws.
H 20cm (8in); S indefinite.

7 *Campanula isophylla* 'Stella Blue' **Campanula**
A compact, trailing campanula with bright blue-violet flowers from midsummer into autumn.
H 15–20cm (6–8in); S to 30cm (12in).

8 *Osteospermum* 'Buttermilk' **Osteospermum**
Daisy-like flowers are primrose-yellow on top and golden-yellow beneath.
H 60cm (24in); S 60cm (24in).

9 *Glechoma hederacea* 'Variegata' **Ground ivy**
A trailing perennial useful in hanging baskets for its attractive, small, toothed leaves with white margins.
H 15cm (6in); S to 2m (6ft).

10 *Viola* 'Jackanapes' **Viola**
Almost all violas and pansies suit hanging baskets. This variety has bicoloured flowers from late spring.
H to 12cm (5in); S to 30cm (12in).

11 *Chlorophytum comosum* 'Vittatum' **Spider plant**
A variegated form of this spreading, dangling, grass-like perennial.
H 15–20cm (6–8in); S 15–30cm (6–12in).

12 *Scaevola aemula* **Fairy fan flower**
Clusters of flowers should appear all summer if plants are fed monthly.
H to 50cm (20in); S to 50cm (20in).

13 *Convolvulus sabatius* **Convolvulus**
A trailing plant whose trumpet-like flowers should spill over the edges of hanging baskets all summer.
H to 15cm (6in); S 50cm (20in).

14 *Helichrysum petiolare* 'Limelight' **Helichrysum**
Trailing, evergreen shrub suited to growing among flowering plants for its lime-green foliage.
H 40cm (16in); S 1.5m (5ft).

15 *Fuchsia* 'Corallina' **Fuchsia**
Its lax, spreading habit makes it ideal for spilling out of a basket. Bright scarlet, purple, and pink flowers.
H 45–60cm (18–24in); S 60–75cm (24–30in).

16 *Nemesia caerulea* **Nemesia**
Spreading, dark green foliage and small, lightly scented flowers, usually purple-blue.
H to 60cm (24in); S 30cm (12in).

INDOOR/OUTDOOR PLANTS

With milder winters, the range of tender plants we can grow is expanding. Plants once thought of as tender may survive outdoors all year in sheltered town gardens or other warm microclimates. Elsewhere, plants can be grown outdoors until temperatures fall, then moved indoors into a cool conservatory or glasshouse. The key to overwintering them is to keep them cool and the soil dry. Once winter is over and the danger of heavy frosts has passed, the pots can go outdoors again. Acclimatize them gradually to outside conditions: if it's cold at night, bring them in, and shelter them from high winds or heavy rain.

1 *Pelargonium* 'Mr Henry Cox' **Pelargonium**
A so-called "fancy-leaved" variety, with golden yellow foliage that has green, purple, and red markings. Almost all pelargoniums need overwintering indoors. Min. 2°C (36°F).
H 25–30cm (10–12in); S 10–12cm (4–5in).

2 *Lantana camara* 'Radiation' **Lantana**
Large, green leaves and round, clustered flowerheads in vibrant orange and red. Needs protection from cold in winter. Min. 10°C (50°F).
H 1–2m (3–6ft); S 1–2m (3–6ft).

3 *Cordyline australis* 'Variegata' **New Zealand cabbage palm**
Palm-like tree with long, arching leaves, striped creamy white at the margin. Mature examples bear tiny, creamy white flowers. Min. 5°C (41°F).
H 3–10m (10–30ft); S 1–4m (3–12ft)

4 *Pericallis* x *hybrida* **Cineraria**
Daisy-like, tender perennial from the Canary Islands, Madeira, and the Azores. A wide range of flower colours, from red and pink to white and blue.
Min. 7°C (45°F).
H 45–60cm (18–24in); S 25–60cm (10–24in).

5 *Ensete ventricosum* **Abyssinian** or
Ethiopian banana
Huge, arching leaves each more than 3m (10ft) in
length. The plant may produce fruits – though not
true bananas. Min. 7°C (45°F).
H 6m (20ft) or more; S to 5m (15ft).

6 *Cycas revoluta* **Japanese sago palm**
Long, arching, deeply divided, feather-like leaves
up to 1.5m (5ft) in length. It may flower, or even fruit,
but it's unusual in container-grown plants.
Min. 7°C (45°F).
H 1–2m (3–6ft) or more; S 1–2m (3–6ft) or more.

7 *Crassula arborescens* **Silver jade plant**
A succulent originally from South Africa, with fleshy,
grey-green leaves and star-shaped, pink flowers in
autumn and winter. Min. 5°C (41°F).
H 4m (12ft); S 2m (6ft).

8 *Echeveria pulvinata* **Plush plant**
A small, Mexican succulent forming rosettes of
fleshy, slightly hairy, green-grey leaves with red
tips and margins. Min. 7°C (45°F).
H 30cm (12in); S 50cm (20in).

9 *Sansevieria trifasciata* 'Golden Hahnii'
Variegated snake plant
A small perennial with rosettes of sharp, pointed,
yellow-and-green striped leaves. In summer, move
outdoors into a sunny position, and do not allow
pots to become waterlogged. Min. 13°C (55°F).
H 12cm (5in); S 12cm (5in).

EDIBLES

Growing your own food is one of the most rewarding aspects of gardening. Not only will you help the environment by reducing the air miles of what you eat, but you'll save money, too. For families, a veg plot in the back garden can also provide a delightful opportunity for children to learn about where food comes from. Your own harvest may not look as uniform as the produce you see in shops, and for good reason: farmed crops are often grown to look "perfect", and anything that doesn't quite make the grade may well end up being disposed of, leading to massive amounts of food waste. While your crops may be a little wonky, remember that you will never buy produce as fresh as the fruit and vegetables you grow yourself.

Nothing can compare with going out to the garden and plucking a perfectly ripe tomato from the vine, or lifting the season's first crop of carrots from the soil.

WHERE TO GROW CROPS

While the idea of growing your own fruit and veg may conjure images of vast beds and huge plots of land dedicated to produce, the reality is much different. It is possible to grow a wide range of crops in a limited space, especially if you use a raised bed. You don't even need to set aside an area of your garden for a dedicated veg patch: most edible plants are attractive in their own right, and can be planted alongside ornamentals to create stunning effects.

FINDING SPACE TO GROW

If you have a limited amount of space for growing vegetables, there are techniques to help you maximize the amont of crops you can plant. Intercropping, for example, involves growing fast-maturing crops among others that take a longer period to mature; for example, by sowing quick-growing lettuce beside rows of Brussels sprouts. This way, you are able to produce a mix of different veg from one small space.

If you don't have space for a dedicated vegetable patch, grow crops among ornamental plants in your borders. Many edible plants are also eye-catching in their own right: Swiss chard, with its striking, deeply veined red foliage, makes an excellent addition to an ornamental bed, while the fine, fern-like foliage of carrots offers a great contrast to broad-leaved plants such as hostas.

Many vegetables can now be bought as dwarf varieties, offering another space-saving solution for small gardens. The likes of Tom Thumb lettuce or "Igloo" cauliflower can be grown easily in limited spaces, such as window boxes or in individual pots (see p. 232–33).

Red chard has eye-catching red stems and red-veined leaves that look lovely in a garden and are delicious to eat.

A selection of cut-and-come-again produce, including lettuce, spinach, and herb plants, provides a constant supply of leafy crops.

Even a small windowsill can be put to good use: lettuce and tomato plants are both good choices for a sunny spot.

THE CLASSIC VEGETABLE PLOT

The traditional approach to growing vegetables is in a dedicated bed in straight rows, as you would see in an allotment (*see right*). Growing crops in this way is very pleasing to the eye, and some would say that a well-planted vegetable plot, with its straight, neat rows, is just as beautiful as any flower border.

A veg patch may be a simple bed dug directly into the ground, or a raised bed filled with good-quality topsoil (*see p.244*). Either way, the width should be no more than 1.2m (4ft), as this allows you to work and reach across the plot easily without too much risk of trampling and compacting the soil. For any plots wider than that, lay down wooden boards to walk across, as this will disperse your weight.

By creating a dedicated space to grow vegetables, you can easily prepare a crop rotation plan (*see p.247*) to ensure that the soil does not deplete in nutrients from one year to the next. Don't forget to make some room for edible flowers, too. The likes of pot marigold and nasturtium not only look beautiful, but can also attract beneficial insects to the plot, providing a natural form of pest control (*see p.325*).

ALLOTMENTS

Having an allotment opens up a whole new world of growing your own crops. If you are lucky enough to acquire a plot of your own, you will find yourself with plenty of space in which to grow a wealth of produce: not only vegetables, but fruit bushes and trees, too (*see pp.262–75*).

While the opportunities afforded by an allotment are great, you should take care not to take on too much all at once; otherwise, your plot may become a full-time occupation. If you are just starting out, consider taking on a half-plot if regulations allow, or putting down turf or wildflower seed (*see pp.180–85*) across the rest of the space until you get the hang of things. If you feel overwhelmed, don't worry about asking for advice: as you spend time in your new plot, you soon realize that there is a community of experienced allotment holders all too ready to share their knowledge.

With a little time and careful planning, a well-run vegetable garden can supply fresh crops all year round.

GROWING VEGETABLES

Before you start out growing your own veg for the first time, familiarize yourself with a few basic techniques. These will not only make the process easier, but will also help improve your chances of a successful crop. Planning is particularly important with a vegetable garden as you'll have to consider crop rotation (*see p.247*), as well as the eventual height and spread of some vegetables, such as cucumbers and runner beans, which will need plenty of room in which to grow.

Raised beds are particularly good for growing crops on poorly drained heavy soil, as the bed will drain better and warm up faster in spring.

PREPARING YOUR SOIL

To some, digging can feel like a chore (especially on a cold day), but it has plenty of benefits. Digging over a veg patch helps to bury and suppress annual weeds, and it is good for incorporating organic matter into the soil, too. Timing your digging will vary according to your soil type:

• **Heavy clay soils** are best dug over in autumn or early winter, to allow the soil to be broken down by rain, snow, and ice later in the season. This ensures the soil is ready for preparing the seedbeds in spring (*see opposite*).

• **Lighter, sandy soils** are best dug over in January or February, as they can collapse and become compact under the weight of winter rain if dug in the autumn. It's best not to do any major digging in summer, when dry weather can take a lot of moisture out of the soil.

Single digging (*see p.294*) is all that is required, leaving the soil ready to sow in spring.

If the soil is shallow or compacted, invest in raised beds at least 15cm (6in) above the surrounding ground, and ideally 30cm (12in), with paths around them wide enough to take a wheelbarrow of 50cm (30in). Fill the beds with well-manured good-quality topsoil. Raised beds need no digging if treading on them can be avoided; instead, apply an annual surface mulch of well-rotted manure or compost. The plants in raised beds get more light and can be planted 30 per cent closer together than they would be if planted directly into the ground, compensating for the area occupied by wide paths.

Rake over seedbeds before sowing to break the soil down to a fine tilth.

Meeting your crops' needs

For gardening purposes, vegetables can be roughly divided into distinct categories, each with their own nutritional needs. Gardeners often use a crop rotation plan to meet these needs and to control root-based pests and diseases. If you're just starting out with a new veg patch, you may want to consider adding one or two extra ingredients if you feel your soil will need it (*see right*).

Brassicas, for example (*see p.255*), need slightly alkaline soil, so you may want to consider adding a little lime to your beds if your soil pH is below 7. Lime is particularly useful if you're going to be sowing into heavy clay soil, as it joins together the tiny clay particles into larger clumps, making it easier to cultivate. Always follow manufacturer's instructions when using lime, and wear gloves and a face mask while handling it.

Preparing seedbeds

After digging over your soil the previous year, by spring you should find that the soil has broken down into small lumps. As the weather begins to warm up, lightly fork over the soil and clear away any weed seedlings. Break down larger lumps of soil with a fork and tread it firm. Rake the soil back and forth, removing any stones as you go, until a fine crumbly soil is achieved. This makes it easier to sow the seeds at the correct depth.

ORGANIC MATTER

There are various forms of organic matter you can use to enrich your soil. Manure, in particular, provides plenty of nutrients to your growing crops. All manures need to be left to rot for a few months before use, particularly if they contain wood chips.

Horse manure, packed with straw, is a highly nutritious option. If you live near a livery stable, ask if you can take some off their hands – they are usually glad to give it to gardeners.

Farmyard manure, made from cow dung, tends to be wetter than horse manure, making it more useful on sandy soils.

Chicken manure is a useful source of nitrogen. It is typically sold as dried pellets.

SOWING VEGETABLES

Raising crops from seed is, in many ways, not unlike growing any other type of plant from seed. How – and where – you choose to sow them depends on how hardy they are and your garden's microclimate.

• **Sowing seeds outdoors** doesn't require any specialist equipment. Seeds are usually sown in shallow trenches ("drills"), in varying ways depending on the seed type. Small seeds, such as carrots, are sown thinly along the drill by taking a pinch of seeds between your forefinger and thumb and trickling them along the row. Larger seeds, like broad beans and peas, are sown as "stations": that is, spaced individually about 15cm (3in) apart in double or triple rows in wide, shallow drills around 2.5cm (1in) deep. Once the seedlings develop, they can be lifted with a trowel ready for their next stage of growth.

• **Sowing seeds indoors** is a good option for more tender vegetables, such as peppers and tomatoes (see pp.258–59). A greenhouse is ideal for sowing seeds, but a small propagator on a windowsill will do nearly as well if need be. Check the seed packets for guidance on the ideal germination temperature, as anything too high or too low may inhibit the seeds.

Alternatively, you can start seeds off in an airing cupboard, but remember to move them out as soon as the seeds begin to germinate and place them somewhere where they will get good light. It doesn't matter what kind of container you use, so long as it has drainage holes. It is always best to use a proprietary seed compost bought fresh when sowing seeds indoors.

(above) **A propagator tray** can be used to start off vegetable seeds indoors on a windowsill.

(below) **Sow small pinches of fine seed** or, as shown here, several larger seeds to ease subsequent thinning.

GROW BAGS

Ideal for first-time gardeners, grow bags are simply sacks filled with fresh compost designed to be planted directly into. While often used for growing tomatoes, they can be used for a range of crops, including salad vegetables, dwarf pea or bean varieties, cabbages, and some root vegetables, including shallots and potatoes. They're great if your space is limited – even a small balcony can hold a grow bag or two – and a fun way of getting children involved in growing their own food.

There is no need to buy specially labelled "grow bags". Simply take a standard 50- or 60-litre bag of compost, pierce a few holes in the base for drainage, then turn it over and cut out a few areas where you can plant up your vegetable seedlings.

Grow bags are a great way of growing vegetables and herbs if you don't have a lot of outdoor space.

CROP ROTATION

To raise vegetable plants successfully, you need to grow them on a fresh piece of ground that has not hosted the same type of crop for at least a year, or preferably three years. This helps to keep down the risk of pests and diseases, and evens out the depletion of different plant nutrients in the soil. If you have several plots, "rotate" them every year following a simple crop rotation plan (*see right*). Alternatively, if you have just one plot, you can divide it into smaller sections and follow the plan on a smaller scale.

• **Plot 1** Dig in compost or manure. Suitable crops include beetroots, carrots, celery, courgettes and marrows, onions, leeks, potatoes, and tomatoes.

• **Plot 2** Use fertilizer when planting or sowing seeds. Suitable crops include beans, chicory, Swiss chard, globe artichokes, lettuce, peas, and spinach.

• **Plot 3** Apply fertilizer before planting each crop. Apply lime if and when necessary and convenient, to keep the soil between pH 6.5 and 7. Suitable crops include broccoli, Brussels sprouts, cabbages, kale, cauliflowers, Chinese cabbages, kohl rabi, radishes, swedes, and turnips.

YEAR 1

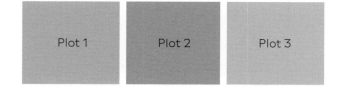

Plot 1 | Plot 2 | Plot 3

YEAR 2

Plot 2 | Plot 3 | Plot 1

YEAR 2

Plot 3 | Plot 1 | Plot 2

Dig manure as needed into the soil each year, and apply lime in winter when the pH falls below 6.5.

Crop rotation delays and limits the build-up of pests and diseases in the soil, resulting in a healthier crop.

ROOT VEGETABLES

As their name suggests, root vegetables are typically grown for their swollen roots or other underground parts. Below, we've covered just a few types of root veg that you can grow quite easily as a beginner. The advice here is quite general; check the seed packet for specific planting and harvesting times.

CARROTS

Look for early cultivars that have long, thin roots. These are ideal for first-time carrot growers, as they need less growing space and mature by midsummer.

• **Sow round varieties,** such as 'Paris Market', indoors in cell trays in early spring, with around 5–6 seeds per cell. Plant outdoors as clumps 15cm (6in) apart under cloches when there are enough roots to grip the soil.

• **Sow outdoors** in any fertile, well-drained soil in full sun or light shade from late March until midsummer. Aim to sow seeds a finger width apart, in rows 30cm (12in) apart. Later, thin seedlings spaced to every 7cm (3in).

• **Harvest** early carrot varieties in midsummer as soon as the roots are a good size. Leave maincrop varieties until autumn when you can protect from frost with straw or store them indoors.

BEETROOTS

Quick and easy to grow, beetroot can vary in colour from deep red to yellow or creamy white, depending on the variety.

• **Sow indoors** in cell trays in early spring, one seed per cell. Each seed produces several seedlings. Plant clumps under cloches 30cm (12in) apart each way, once roots are sufficient to bind the potting compost.

• **Sow outdoors** in open ground in an area of full sun from late March to midsummer. Scatter in shallow drill stations no more than 2.5cm (1in) deep, in clusters of two or three seeds at 7.5cm (3in) apart. Once the leaves are large enough to handle, thin out plants to leave one seedling per station.

• **Harvest** from summer to autumn, once the roots reach around the size of a golf or tennis ball, depending on the variety.

PARSNIPS

Parsnips are an excellent winter vegetable. They are easy to grow, but require a long growing period and seed germination can be erratic.

• **Sow outdoors** in shallow drills in a sunny spot in fertile, well-drained soil. Sow 2–3 seeds per station, with each station spaced 30cm (12in) apart. For smaller roots, space closer at 10cm (4in) apart. Thin seedlings once the seed leaves are large enough to handle, thinning to one seedling per station. Seedlings don't transplant.

• **Harvest** from autumn through winter. The flavour of the vegetable is improved after frost.

RADISHES

Radishes are a quick-growing crop, good for intercropping (see p.242) with other crops that take longer to mature. Ideal for children to grow.

• **Sow indoors** in large pots in late winter or early spring, with seeds a finger width apart. Cover with sieved compost and water. Later, thin to one plant every 5cm (2in). Place pots outdoors in a greenhouse or sheltered place once seedlings emerge.

• **Sow outdoors** in open ground anytime from spring to autumn, scattering seed along a row. As the seedlings develop, thin them out to leave plants spaced 5cm (2in) apart. Radishes don't transplant.

• **Harvest** regularly while the roots are young. For a continuous crop, sow short new rows at 10-day intervals throughout the growing season, spacing the rows 15cm (6in) apart.

(opposite, clockwise from top left) **Carrots, beetroots, parsnips, radishes**

STORING THE HARVEST

Many root vegetables, including carrots, beetroot, parsnips, onions, and potatoes, can be lifted and stored over winter in a cool, dry, frost-free place. Regularly inspect them for signs of rot, disposing any that are affected before it can spread to the other roots.

Alternatively, if you have an area of sheltered, well-drained soil to spare over winter, you could try clamping your potatoes. Cover the area with a 20cm (8in) layer of straw, and arrange the potatoes in a pyramid shape on top. Cover this pyramid completely, with a 20cm (8in) layer of straw, then top with a 15cm (6in) layer of patted-down soil, leaving a side "chimney" (wrist-width tuft of straw) sticking through the soil covering to allow moisture to escape. Carrots and beetroots can also be clamped but there is no need for straw.

As well as enjoying the roots, save the young leaves of kohl rabi plants and add them to summer salads.

(top) **Clamping potatoes in straw** is an unusual yet effective way of storing them over the winter.

(bottom) **Layer carrots** in a sturdy box and cover each layer with compost to keep them through the winter.

KOHL RABI

While grown for their attractive roots, kohl rabi are in fact a member of the brassica family (*see p.255*), and can be treated as such in a crop rotation plan.

• **Sow outdoors** into seedbeds in a sunny spot between spring and midsummer. When the seedlings are large enough to handle, transplant them to their final growing position: the plants should be spaced 23cm (9in) apart, in rows spaced 30cm (12in) apart.

• **Harvest** when the roots are about the size of a tennis ball or slightly smaller. They can be stored for a short time.

(above left) **Multi-seeded trays** are a good way to grow onions from seed indoors. Plant out as clumps from mid-spring.

(above right) **Leave onions** to dry for two weeks in a well-ventilated shed or greenhouse before storing.

ONIONS

A mainstay of the vegetable garden, onions are one of the most versatile crops you can grow. They can be raised from seed, but if you're new to growing your own veg, the best method is to plant onion "sets": tiny, immature onions that are widely available in garden centres and via mail order.

• **Sow seed indoors** in midwinter for an early onion crop. Sow 4–5 seeds per cell. Plant out in spring, treating them as you would sets (*see below*).

• **Plant sets outdoors** in spring into a sunny, well-drained area about 15cm (6in) apart, in rows spaced 30cm (12in) apart. Gently push each set into the soil so that only the tip remains showing, then firm the soil around them.

• **Harvest** as soon as bulbs are usable in early summer (for plants started off from seed indoors) or late summer to early autumn (for plants grown from sets). Once the foliage starts to turn yellow and topple over, lift the bulbs. Dry them on a rack outside in full sun, in a well-ventilated greenhouse or shed, or under cloches. Leave them for about two weeks to allow the onions to develop dry, rustling foliage. They will then keep for months in a cool, dry space.

THE ALLIUM FAMILY

Onions are perhaps the most well-known of the allium family, which contains a number of popular vegetables.

Shallots are generally quite like onions in how they grow, but are usually hardier, meaning that they can be planted in autumn. They can be rasied from seeds sown in spring, but sets are easier.

Spring onions are fast-growing, and are a great option for intercropping. Sow successively every two weeks throughout summer, in shallow drills around 2cm (¾in) apart with around 1cm (½in) between each seed, so that there is no need to thin them out later. Space rows about 15cm (6in) apart. If you sow the seeds thinly there is no need to thin them out later. They should be ready to harvest 4–6 weeks after sowing.

Garlic is an essential ingredient in the kitchen that is surprisingly easy to grow. In spring, take a head of garlic, break it into individual cloves, and plant into well-drained soil in a sunny spot, about 5cm (2in) deep and 15cm (6in) apart. Harvest in early autumn or late winter.

Shallots grown from sets form clumps that mature into clusters of dry bulbs.

POTATOES

As root vegetables go, potatoes are an undisputed staple. There is nothing quite like harvesting the first early potatoes of the season, and seeing those creamy white tubers appear as you dig them up. Potatoes can be grown in several different ways, but typically involve starting with small seed potatoes, available from most garden centres, which have been allowed to "chit" (*see right*) ahead of the growing season.

• **Plant earlies outdoors** from early to mid-spring, 8–15cm (3–6in) deep. Space the seed potatoes 30–39cm (12–15in) apart in rows 39–50cm (15–20in) apart. As the shoots emerge, watch out for frosts, covering them with horticultural fleece to protect the young growths. You can also protect young plants by "earthing up", or drawing up soil around them, on chilly nights – even a little soil excludes frost. Continue to earth up the plants as they grow, as this prevents young potatoes near the surface from turning green.

• **Plant maincrop varieties outdoors** from mid- to late spring 8–15cm (3–6in) deep. Space the seed potatoes 39cm (15in) apart in rows 75cm (30in) apart. Maincrop potatoes take up more space than early varieties. Keep well watered throughout the growing season.

• **Harvest** early varieties in summer, when they start flowering. These are best eaten fresh, and will only store for a short time. Harvest maincrop varieties on a dry day in autumn, taking care not to damage any of the tubers as you lift. Store in hessian sacks in a dark, frost-free place, or clamp (*see p.250*). If the tubers turn green, they will be inedible.

(opposite, clockwise from top left)
Maincrop potatoes can be particularly space-hungry, so make sure to plan for their eventual size in your vegetable plot.

Protect young potato plants with a fleece.

Prevent tubers greening in sunlight by covering the roots with soil or mulch.

Dig out the tubers for storage once the tops die off in late summer.

CHITTING POTATOES

Potatoes are propagated through a method known as chitting, or sprouting. To chit your own seed potatoes in late winter for planting in spring, position small tubers in egg boxes or a seed tray with their "rose" end (the end with the most "eyes", or buds) facing up. Place the egg boxes or tray in a light, frost-free place indoors, and leave for a few weeks until small shoots begin to emerge.

(above) **Small potatoes** are best for chitting.

(below) **Place small potatoes** in cardboard egg boxes with the majority of their "buds" facing upwards.

LEAFY VEGETABLES

There's nothing quite like the taste of freshly picked salad leaves in a light summer dish, and by growing your own lettuce and rocket in your back garden, you will have a constant supply for weeks on end. As the seasons change and winter draws in, leafy brassicas bring with them their own burst of nutrients to see you through to spring.

ROCKET

This spicy leaf is ideal for mixing with the other salad crops on this page. It should be grouped with brassicas in any crop rotation plan (see p.247).
• **Sow outdoors** at four-week intervals from mid-spring to midsummer. Once large enough to handle, thin seedlings to about 23cm (9in) apart in rows 23cm (9in) apart.
• **Harvest rocket** as soon as the leaves are ready, usually after about two weeks, taking just what you need each time to avoid waste. Discard plants once they flower. Grow autumn and winter crops in a greenhouse or under cloches.

SPINACH

Fast-growing spinach is known for its nutritious leaves, which are a source of iron.
• **Sow outdoors** in early autumn or from early spring until midsummer, in shallow drills 30cm (12in) apart. Thin seedlings to 15cm (6in) apart.
• **Harvest** as soon as the leaves are usable. Discard plants once they flower. Continue to sow seeds at three-week intervals until midsummer for a regular crop.

CHICORY

This bitter leafy vegetable comes in three main forms: green, lettuce-like chicory; red chicory (also known as radicchio), and forcing varieties like 'Witloof' that need to be blanched.
• **Sow outdoors** from early to midsummer in shallow drills 30cm (12in) apart. Once large enough to handle, thin seedlings to 23cm (9in) apart.
• **Blanch forcing varieties** in November. Lift the chicory roots, cut back the leaves to about 2.5cm (1in) from the crown, and replant five roots vertically in a 23cm (9in) pot filled with moist sand or potting compost, leaving the crown unburied. Cover with another pot or a black polythene bag supported with canes; the aim is to exclude all light and keep the plants at around 10°C (50°F).
• **Harvest** green chicory and radicchio from autumn until early winter, and forced chicory from early winter onwards, once they reach around 15cm (6in) high.

LETTUCE

A great choice for intercropping (see p.242), lettuce is an ideal cut-and-come-again salad crop.
• **Sow outdoors** from mid-spring to summer. Continue to sow at regular three-week intervals for a continual crop througout the summer.
• **Harvest** 90–120 days after sowing, depending on variety. If growing a cut-and-come-again variety, sow every two weeks. Take just what you need each time according to the seed packet, returning later for more lettuce leaves, to minimize waste.

BRASSICAS

Many leafy vegetables belong to the brassica family, and should be grouped together in crop rotation plans.

1 Brussels sprouts Sow in pots or cell trays in mid-spring. When the seedlings are about 7.5cm (3in) tall, transplant to their final growing position, spacing them 75cm–1m (2.5–3ft) apart each way. Keep well watered, and stake plants in exposed areas. Harvest from late autumn onwards, from the bottom of the stem upwards.

2 Cauliflower Sow summer varieties in pots or cell trays from mid- to late spring. Plant out when the seedlings are large enough, 23cm (9in) apart each way. Keep cauliflowers well-watered. Harvest 16–40 weeks after sowing, depending on variety.

3 Sprouting broccoli Sow in a seed tray under fleece in late spring. Transplant seedlings when around 7.5cm (3in) high, to their growing positions, spacing them 75cm (2.5ft) apart. Harvest regularly from late winter to ensure continuity of the crop. Support with stakes in exposed areas and earth up stems to increase stability. Harvest 8–12 months after sowing.

4 Cabbage Sow early varieties indoors in early spring. Transplant under cloches when the plants are around 5cm (2in) high, spacing them 45cm (1.5ft) each way. Sow later varieties outdoors under fleece at regular intervals and transplant as above. Harvest early varieties in summer, and later varieties through to late winter.

5 Kale Sow seeds under cloches and fleece from mid-spring and continue sowing outdoors at regular intervals until late spring or early summer. Plant out when the seedlings are about 7.5cm (3in) tall, spacing them 45cm (18in) apart each way. Some varieties can grow to about 90cm (3ft) and will need staking in exposed gardens. Harvest from around 7 weeks onwards after sowing.

6 Pak choi and Chinese cabbage Sow seeds from early summer to midsummer in shallow drills 23cm (9in) apart, sowing two or three seeds at each station. Thin to single seedlings when the plants are large enough to handle. Harvest 8–10 weeks after sowing.

PEAS AND BEANS

The flavour of freshly picked, home-grown peas and beans surpasses anything you can buy in a supermarket. Legumes can take up a lot of space, particularly climbing varieties, so make sure you provide plenty of support for their twining stems and tendrils (*see opposite*). Dwarf varieties may be more suitable for smaller gardens.

PEAS

There are different varieties of peas, including classic garden peas and mangetout, where the whole pod is eaten. All require support as they grow (*see opposite*).

- **Sow indoors** in early spring. Drill drainage holes into a length of guttering, then sow seeds into it, spacing them about 5cm (2in) apart in a double row. Plant out when roots bind to the compost by simply sliding the plants from the guttering into a shallow trench.
- **Sow outdoors** from mid-spring onwards. Scatter seed in a trench around 5cm (2in) deep and 100cm (4in) wide, spacing the seeds about 5cm (2in) apart. Mice love pea seeds, so cover the rows with fine wire netting until they germinate.
- **Harvest** early varieties 11–12 weeks after sowing, and maincrop varieties 13–14 weeks after sowing.

FRENCH BEANS

More delicate in texture than runner beans, French beans are available in both climbing and dwarf varieties. While both are grown in a similar way, climbing French beans can grow up to 2.1–2.4m (7–8ft) tall, and require plenty of support (*see opposite*). Dwarf varieties are ideal if you are short on space, and they can be grown in containers.

- **In colder areas, start seeds off indoors** in cell trays or individual pots in late spring, with one seed per cell or pot. Plant outside once the threat of frost has past.
- **In warmer areas, sow outdoors** in late spring, once the threat of frost is passed. Put three seeds (or seedlings, if started off indoors) at the base of each cane of a tripod. This ensures that, if one fails to germinate, there is another one to take its place. Keep them well-watered throughout the growing season.
- **Harvest** from midsummer onwards, when the pods are about 10–15cm (4–6in) long.

Dwarf varieties of runner beans are ideal if space is limited. Sow seeds 15cm (3in) apart with 30cm (12in) between rows for a heavy crop.

Runner beans can be grown in large containers. They need a lot of water during the growing season to crop well.

RUNNER BEANS

Like French beans, runner beans come in both climbing and dwarf varieties. Like perennial plants, runner beans should be treated as annuals. The flowers are attractive and are shades of red and pink through to white, depending on the variety.

• **In colder areas, start seeds off indoors** in cell trays or individual pots in late spring, with one seed per cell or pot. Keep on a windowsill or in a propagator until the threat of frost has passed.

• **In warmer areas, sow or plant outdoors** as for French beans (*see opposite*), with 2.1–2.4m (7–8ft) tall canes to support their growth.

• **Harvest** the pods when young and tender, around 15cm (6in) long. Pods that are too long become stringy and coarse.

SUPPORTING CLIMBING CROPS

A number of vegetable plants require a frame to support their climbing stems. There are a few different methods of supporting crops, depending on how tall they grow or how heavy they become.

Pea plants only need a few twiggy sticks, about 60–100cm (24–39in) tall, inserted along each row. Alternatively, use well-staked wire netting above the plants, so that the tendrils can cling to it for support.

Climbing beans are often grown on tripods made up of up to 8 canes tied together at the tops (*see below*). Or, make a double row of 1.8m (6ft) canes tied together, with a third, horizontal cane secured at head height. Stems may need tying in to encourage wayward shoots to twine around canes.

Fruiting vegetables, including summer squashes and cucumbers (*see pp.258–60*), require more substantial supports. A sturdy, well-made tripod may suffice; or, for a more decorative effect, try training your crops up a small arbour or pergola-like structure.

If space is limited, many crops can also be trained against a trellis on a wall. This approach also shelters plants from wind, and can make harvesting easier.

FRUITING VEGETABLES

Technically classed as fruits, these vegetables burst with the taste of summer. Fruiting vegetables are tender plants, which means they will not withstand frost, so seeds should generally be started off indoors in mid-spring, or outdoors in milder regions if they are protected with a cloche. For the best crops in colder regions, grow plants in a greenhouse.

Pinch out side shoots as tomato plants develop in order to encourage as much energy as possible into the growing fruits.

TOMATOES

Quintessential summer vegetables, tomatoes are easily grown from seed indoors, or outdoors under cover in milder regions.

They can be divided into cordon (climbing) varieties, which need to be supported (see p.257), and more compact bush varieties.

• **Sow indoors** in early spring into seed trays filled with fresh seed compost, or buy plants in mid-spring. Plant out once the threat of frost has passed.

• **Sow outdoors** under cloches in mid-spring, removing the cloches once the frosts have passed. Keep the plants well-watered throughout the summer months and feed with a high-potash fertilizer according to the manufacturer's instructions. Support cordon varieties as they grow.

• **Harvest** when the tomatoes are fully coloured. If any green tomatoes remain, turn them into chutney.

AUBERGINES

Also known as eggplants, aubergines produce attractive black, white, or red fruits depending on the variety.

• **Sow indoors** in early spring into seed trays filled with fresh seed compost. Plant out into fertile, well-drained soil once the threat of frost has passed. Keep the plants under cloches and well-watered throughout the season. When the fruits begin to form, start feeding with a high potash fertilizer. Thin out fruits to 3–4 aubergines per plant to encourage large fruits to develop.

• **Harvest** from midsummer onwards when the fruits reach full size.

PEPPERS

Sweet peppers ripen to shades of red or yellow depending on the variety. They can also be used as green peppers if picked before they fully ripen.
• **Sow indoors** in early spring into seed trays filled with fresh seed compost. Plant out once the threat of frost has passed but keep under cloches if possible. Water and feed regularly as for tomatoes.
• **Harvest** from around 20 weeks after sowing, either when green or once they have fully coloured, depending on preference.

COURGETTES AND MARROWS

What is the difference between courgettes and marrows? Only the size of the fruits! Courgettes are harvested when they are quite small and tender to eat; if left to grow larger and develop tougher skins, they then become marrows. All the flowers can also be eaten.
• **Sow indoors** under cover in mid-spring, placing two seeds in small pots 2.5cm (1in) deep. Grow the plants on indoors, repotting them into larger containers if they become root-bound.
• **Harden off** and plant outside once the threat of frost has passed. Courgette plants are quite vigorous, so keep them 90cm–1.2m (3–4 ft) apart from one

another. Keep the plants well-watered throughout the summer and autumn.
• **Harvest** courgettes when they are around 10cm (4in) long, picking young fruits little and often to encourage further growth. For marrows, leave fruits to develop, harvesting at whatever size you prefer. If you want to store the courgettes for the winter, harvest once the skin is hard.

CUCUMBERS

Cucumbers may be grown in greenhouses, while hardier varieties (including gherkins) can be grown outdoors.
• **Sow indoors** in mid-spring into seed trays or pots filled with fresh seed compost. If sowing in pots, sow three seeds per pot, then thin out the weaker seedlings before planting outdoors in early summer. Support plants with a 4ft (1.2m) tripod. Water and feed as for tomatoes. Do not remove male flowers as pollination is essential.
• **Harvest** around 12 weeks after sowing, once the fruits reach around 15cm (6in) long.

Don't discard courgette flowers: they can be stuffed, fried, and eaten as part of a summer meal.

In greenhouses use all-female cucumbers. Traditional varieties produce male flowers that should be removed as fertilized fruits are inedible.

PUMPKINS AND WINTER SQUASHES

Pumpkins and winter squashes require a long growing season and plenty of space in order to mature.

• **Sow indoors** under cover in mid-spring, with two seeds per small pot placed about 2.5cm (1in) deep. Remove the weaker seedling to allow the stronger one to grow on.

• **Harden off the seedlings** and plant out once the threat of frost has passed. Position them 1–1.5m (3–5ft) apart in a shaded spot, in fertile soil with plenty of organic matter added. (If space is limited in your garden, try planting a seedling on top of a compost heap; the warm, moist conditions here suit them very well.) Water plants well through summer and feed with a high potash fertilizer every 10–14 days.

• **Harvest** the fruits before they grow too large if you intend to eat them. Otherwise, for Halloween decorations, leave them to early October to allow them turn a beautiful golden colour.

SUMMER SQUASHES

Similar to courgettes in taste, summer squashes come in a wide range of sizes, shapes, and colours. Bushy and trailing varieties are available; trailing varieties should be grown with supports (*see p.257*), as otherwise they can take up a lot of space.

• **Sow indoors** under cover in mid-spring as for pumpkins (*see left*). In milder regions seeds can be sown outside under cloches in mid-spring.

• **Harden off the seedlings** and plant out into moist, rich soil once the threat of frost has passed. Plant bushy varieties 45cm (18in) apart, and trailing ones 90cm–1.2m (3–4ft) apart. Water plants well through the growing season and feed regularly with a high potash fertilizer.

• **Harvest** the fruits while still young, cutting the stems near the base. Take care when harvesting, as some older stems have sharp bristles.

To grow a really huge pumpkin, choose a large fruiting variety.

Summer squashes are especially delicious when stuffed and baked.

GROWING HERBS

Herbs make a wonderful addition to any garden. As well as being valued for both their culinary and medicinal uses, they are very attractive plants in their own right, with wonderful foliage that grows in all shapes, sizes, and textures, and their palette of scents is one of the greatest joys of gardening. If you can, plant them somewhere that you'll walk past regularly, so that they fragrance the air as you gently brush past. They are generally easy to grow, too.

WHERE TO GROW HERBS

If you have the space, consider creating a a dedicated herb garden. While it does not have to be a formal design, let yourself get creative with their different forms to make an attractive feature. If you can, position your herb garden near to the house, so that you can easily pop outside and take a few cuttings before you begin to cook.

If you don't have space for a herb garden, you can still create an eye-catching display by planting up a few favourite varieties in pots with well-drained compost. Try grouping a few together in a window box, such as parsley and basil.

Herbs also make a great addition to an ornamental display. Purple or variegated varieties of sage, for instance, pair brilliantly with dark-leaved *Physocarpus*, while chives make an excellent choice for edging walkways and can be added to many culinary dishes.

Plant herbs close to walkways and high-traffic areas, where their scent can be enjoyed as you pass by.

CULINARY HERBS

1 *Angelica archangelica* **Angelica**
This giant of a herb is a short-lived perennial.
Sow seeds in early autumn where they are to grow,
or sow in pots and plant out when large enough.
Grow in a sheltered spot and water regularly
throughout the summer. The flower heads appear
after several years.

2 *Melissa officinalis* 'Aurea' **Lemon balm**
This herb tolerates dry soils and has a lovely lemon
scent when the leaves are crushed. It grows well in
pots, and seeds can be sown outdoors from April
onwards. Cut it back regularly to promote new growth.

3 *Ocimum basilicum* **Basil**
This tender annual is best started off indoors and
planted out when the threat of frost has passed. Sow
indoors in mid-spring. Pinch out the growing tips
regularly to encourage more growth.

4 *Laurus nobilis* **Bay**
A very popular evergreen shrub; its leaves make an
excellent addition to sauces and fish dishes. It is often
grown in containers and trimmed into various shapes.
Best bought as a pot-grown specimen and placed in
a sheltered spot, away from cold winds.

5 *Borago officinalis* **Borage**
This bee-friendly herb offers edible leaves, stalks, and
delicate blue flowers, which bloom in summer. Sow
seeds in spring or summer outdoors, either in the
ground or in pots, thinning them out as they develop.
Borage has a long tap root so dislikes transplanting.
Plants grow to about 1m (3ft) tall.

6 *Allium schoenoprasum* **Chives**
This allium (*see p.251*) offers mild-flavoured stems
that can be cut and used in cooking all through
summer. The flowers are edible, too. Sow seeds in
spring, or buy pot-grown clumps, spacing them 23cm
(9in) apart. Divide clumps every 3–4 years. Trim the
stems to about 2.5cm (1in) from the soil.

7 *Foeniculum vulgare* **Fennel**
An attractive perennial herb with fine, feathery
foliage. It is best grown at the back of a border
rather than in a herb garden, in a sunny spot with
well-drained soil. Sow seeds in spring, or buy plants
from a herb nursery.

8 *Mentha* spp. **Mint**
With so many different varieties to choose from, this
is one of the most popular herbs to grow. Mint can
be invasive, so it is better to grow it in pots with the
rim of the pot above ground to prevent it spreading.

9 *Origanum onites* **Marjoram**
A half-hardy herb, marjoram seed can be sown indoors in spring and planted out when the threat of frost has passed. Space plants 23cm (9in) apart in well-drained soil in a sunny spot. In autumn, plants can be potted up and taken indoors for a winter supply. Oregano (*O. vulgare*) is a close relative.

10 *Petroselinum crispum* **Parsley**
There are several varieties and all taste delicious. Sow seeds indoors in mid-spring for a summer crop. Germination can be slow. Sow again in autumn, and grow in containers indoors for a crop through winter.

11 *Salvia rosmarinus* **Rosemary**
Formerly known as *Rosemarinus officinalis*, rosemary makes an attractive shrub in any border. Buy as pot-grown plants in early spring, then transplant them into well-drained soil in a sheltered spot. Harvest young growth regularly to provide plenty of tender shoots. Rosemary is also easy to grow from cuttings.

12 *Salvia officinalis* **Sage**
The sage plant's aromatic, grey–green leaves and blue flowers are a wonderful addition to a garden. One plant is usually enough. Plant a pot-grown one in spring and trim or harvest shoots regularly to keep the plant bushy.

13 *Rumex acetosa* **Sorrel**
A broadleaved herb that can grow to 45cm (18in) in height. The French variety (*R. scutatus*) has smaller leaves. Plant seeds or roots outdoors in autumn. Sorrel prefers partial shade and moist soil enriched with plenty of organic matter.

14 *Artemisia dracunculus* **Tarragon**
Tarragon has a spreading growth habit, similar to mint. In early spring, plant pot-grown specimens in a sheltered spot with well-drained soil. Pick leaves from early summer to autumn.

15 *Coriandrum sativum* **Coriander**
An annual herb with divided foliage and small white or mauve flowers. As well as picking the young leaves, you can also gather seeds from seed heads as they ripen.

16 *Thymus pulegioides* 'Aureus' **Lemon thyme**
Less pungent than common thyme, this herb has a wonderful citrus flavour. Plant pot-grown specimens in spring in well-drained soil 30cm (12in) apart. Divide every three or four years. Lemon thyme also grows well in pots and on windowsills.

GROWING FRUIT

One of gardening's greatest pleasures has to be growing your own fruit. There's nothing quite like sitting under an apple tree after a day of hard work – and with modern advances in dwarfing rootstocks, the possibility of a home-grown apple crop is now a real option for more and more gardeners who might have previously thought that they would not have enough space. In small gardens, you can still grow a variety of bush fruits, such as strawberries, in pots or beds. Many fruit varieties require at least two different plants in close proximity in order to produce fruit, although some are self-fertile; always check before buying whether one or two plants will be needed.

While a full orchard may feel out of reach to most gardeners, modern dwarf varieties allow even smaller gardens to make room for a few fruit trees.

CHOOSING WHAT TO GROW

Thanks to the development of modern dwarfing rootstocks and the availability of self-fertile plants, most people can grow fruit trees in even the smallest of gardens. Many fruit tree varieties can be highly ornamental, with beautiful spring blossom, making them a stunning – and productive – addition to an ornamental border.

If you're not quite ready to try growing a fruit tree or two, fruit bushes provide a cheaper, more compact alternative. Soft fruits, such as redcurrants, white currants, gooseberries, and blackcurrants can all be grown in ornamental borders, while strawberry plants can easily be grown in containers.

The area you live in will to some extent dictate the kind of fruit trees and bushes you can grow outdoors. For example, it would be difficult to grow peaches outdoors in the north of Scotland. These would need some protection under glass in such a cold, wet climate. Frost-prone areas, such as deep valleys, are also generally unsuitable for growing fruit trees, given the risk of frost damaging the flowers. If your garden is on a slope, plant fruit trees and bushes at or near the top of the slope, as cold air always falls to the lowest point.

PLANTING FRUIT BUSHES

If you don't quite have space for a fruit tree or two, fruit bushes provide an excellent alternative – and the flavour of freshly picked soft fruits and berries surpasses anything that can be bought in a punnet in a supermarket.

While fruit bushes do have some care differences, the advice for planting them is generally the same. Prepare the soil well in advance of planting by incorporating plenty of well-rotted organic matter. For plants bought in pots, give the plants a good soak while still in their containers, then dig a hole slightly larger than the rootball and set the plant inside, so that the rootball is slightly below soil level. Firm the soil around the roots and water in well. Bareroot fruit bushes are planted during their dormant season, from November to March. Dig a hole large enough to allow the roots to spread out, then place in the hole, firm the soil around the roots, and water in well.

Blackcurrant bushes (*see p.275*) are the exception to the guidance above. These plants need to be planted with around 5–8cm (2–3in) of their stems buried beneath soil level, to encourage new shoots to grow from the base.

CARING FOR FRUIT BUSHES

Newly planted fruit bushes will need to be kept watered during dry spells, especially in their first year while the roots become established in the soil. Keep weeds at bay, too; not only do they compete for moisture in the soil, but they often harbour pests and diseases, which can affect the bushes' development.

Some bushes, such as raspberries, redcurrants, and whitecurrants, will require pruning to encourage fruit growth and allow air to circulate between the stems, reducing the risk of disease.

(top right) **Even a small corner of a garden** can produce a bountiful fruit crop, with berry bushes and even a miniature apple tree planted in repurposed containers with drainage holes.

(below right) **Many berry and currant bushes** require some pruning each year in order to encourage the development of future fruits.

PLANTING FRUIT TREES

Much of the work involved in planting and caring for fruit trees is the same as for ornamental trees. They can be bought as bare-root or container specimens, and should be planted accordingly (see pp.197–99). Bare-root trees can be planted from November to March, while container-grown trees can be planted at any time of year, but ideally in autumn or winter.

Before planting, improve the area where the tree is to be planted with plenty of organic matter, as this will help retain moisture and improve the soil. If planting a grafted tree, make sure that the point of grafting (which is easily identified as a swollen part low down on the trunk) is kept well above ground level. If it is at or below ground level, the upper, grafted part of the tree may take root, and the benefits of the rootstock will be lost.

If planting your tree into a lawn, leave an area 1–2m (3–6ft) around the tree free of grass, as the grass will compete for nutrients and moisture. After 3–4 years, once the trees have been established, the grass can be allowed to grow back.

Tree fruits are ready to harvest when they come away easily after being gently twisted from the branch.

Young trees (around one year old) may take 3–4 years before fruiting for the first time, while mature specimens may produce fruits the first year after planting. Grafting can also affect the vigour of the tree and, therefore, when you can expect your first fruits. On the other hand, semi-mature container-grown trees may be planted with fruits already on them.

PRUNING AND CARING FOR FRUIT TREES

It's tempting to wonder why we need to go to all the bother of pruning fruit trees and giving them the care they need to produce a good crop. If you encounter an unpruned fruit tree, you might notice, at first, that it produces far more fruit than a pruned tree grown in similar conditions. These fruits, however, are likely to be small, and because they are overcrowded on the stem, they are far more likely to be damaged, leading them to rot quickly. So, while a pruned tree may produce fewer fruits overall, what it does produce will be more regular, larger, and generally of a better quality.

There are two basic principles to follow when pruning fruit trees:

• **Pruning in summer** encourages more fruit to develop. In midsummer, prune all side shoots back to four or five buds from the point of growth; or, if you have a standard tree (see pp.201–05), cut back the current year's growth to three or four buds and keep the centre of the tree open. This pruning encourages the formation of "spurs", which are small growths that produce fruit buds.

• **Pruning in winter** encourages vigorous regrowth. Generally, the harder you prune, the more growth your tree will put on. In the first few years, prune each winter to form the shape you want. Once you achieve the size and shape you want, winter pruning should only be necessary if there are overcrowded branches, crossing branches rubbing against each other, or dead or diseased branches to be removed.

Thinning

This is the process of removing excess or weaker fruits early in their development in order to focus a tree's energy into the rest of the crop. Apple trees do this naturally: you may have heard of "June drop", when smaller fruits are shed from the tree.

Prune fruit trees in winter as you would for any other type of tree, making sure to remove dead, diseased, and crossing branches as necessary.

You may also need to thin fruit trees yourself if you want good-sized fruits. It is usually enough to remove the "king fruit" (in the centre of a cluster), as this will redirect energy to the rest of the cluster. You can also thin out developing fruits further, leaving remaining fruits around 5–8cm (2–3in) apart, to futher encourage them to grow to their full potential.

Encouraging new growth

As the years progress, you might notice a few bare, unproductive branches on your fruit tree. To allow dormant buds to break, a method known as "notching" can be tried. Immediately above a latent bud, score a small line into the hard wood with a sharp knife and remove a crescent-shaped piece of bark. This releases the bud from dormancy by encouraging a flow of hormones within the plant that enables the dormant bud to grow. Where this fails, there is little else to do but to remove the branch, which will promote the formation of new shoots to replace it.

HARVESTING

Fruit skins bruise easily, causing them to rot quickly, so handle with care when harvesting and do not store any fruits that are damaged. Apples and pears are usually ready to harvest when you carefully lift and twist the fruits and they come away easily.

MAINTAINING FRUIT TREES AND BUSHES

In order to achieve the maximum yield of good-quality fruits, it is vital that you care for your fruit trees and bushes from year to year.

• **Water fruit trees and bushes well,** especially while young trees are becoming established, and when fruits are developing. Make sure to give them a good soaking once a week, rather than watering them little and often. Small amounts of water every day encourages roots to grow towards the surface of the soil, which can then make them more vulnerable to dry conditions.

• **Feed fruit trees and bushes** at least once a year to achieve the best yields. A potassium-rich general fertilizer is a good organic option; otherwise, use a general feed high in potash, such as rose fertilizer. Apply in later winter or early spring. Don't forget to mulch, too (*see p.200*).

• **Keep weeds at bay,** especially around newly planted trees and bushes. Weeds compete for moisture and nutrients, and can also harbour pests and diseases. Hoe off weeds regularly, or place a piece of polythene around the base and cover it with a little soil or other organic matter to suppress weed growth.

• **Spray trees and bushes** only when it is needed, as soon as you spot any pests or signs of disease. Always follow the manufacturer's instructions and spray in the evenings, after bees and other pollinators have gone for the day. If you would prefer to care for the trees organically, do all you can to encourage natural predators and establish a biodiverse garden (*see p.325*).

Water container-grown trees well, particularly when you see the buds of young fruits begin to develop.

FRUIT TREES

You don't need an orchard to grow your own tree fruits. With just a couple of trees, you can produce a delicious crop year after year, even from specimens grown in containers. Many fruit trees are sold grafted onto rootstocks, which helps to control the tree's eventual height and spread. General information about rootstocks has been provided below; for more detailed information, contact a specialist fruit grower.

APPLES

There is a bewildering array of apple varieties to choose from. Consider which varieties produce the best flavour, and which ones offer a long harvesting period. Ask local growers, allotment holders, or garden centre staff which varieties are popular in your area, as that will give you a good idea of what will succeed in your garden. Remember, that in order to crop successfully, apples need to be pollinated by a different, compatible variety. Check with a specialist fruit tree grower for varieties that are suitable for cross-pollinating the tree you have chosen.

Rootstocks

A wide range of apple tree rootstocks are available to home gardeners, with three being particularly notable. M27 rootstocks produce the smallest trees, roughly 1.5–2m (5–6ft), while M9 trees will be slightly larger, at approximately 2–2.5M (6–8ft). Both of these can be planted into containers, or into the ground if you have rich, fertile soil. Be aware, though, that trees with these rootstocks will probably need staking for most of their lives. If your garden soil is of poorer quality, MM106 rootstock is a good all-rounder. It is also slightly more vigorous than M27 and M9 rootstocks, growing to approximately 3–4m (10–12ft). Always check the specific height and spread of your chosen variety, and remember that soil type and planting location can also affect a tree's eventual size.

Harvest and storage

Early varieties are best eaten soon after harvesting in late summer, although they can be kept for a few weeks if kept refrigerated. Later varieties, which are harvested in autumn, can be stored well into winter, or even into the following year, if stored correctly. Store small quantities in sealed polythene bags, keeping different varieties in separate bags. Prick a few holes in each bag to allow a little air in and keep in a cold, frost-free place. Only store undamaged apples; damaged fruits will rot quickly and affect other apples. Inspect them regularly and remove any showing signs of rotting.

Apples that ripen on the tree are best eaten immediately.

PEARS

Pears are a little more difficult to grow than apples as they generally flower earlier, in early to mid-spring. This makes them more susceptible to frost damage, which can affect the fruit crop later in the year. For this reason, pear trees are best grown in a sheltered spot, or trained against a wall or fence. For further protection, if the tree is hit with very cold conditions while in flower, drape horticultural fleece over the branches.

Compared to apples, there aren't as many varieties of pear tree, but the same rules about pollination apply (*see opposite*). One pear variety, 'Conference', is self-fertile to some extent and will produce a decent crop without the need for cross-pollination, although for best results it is preferable to have another variety planted nearby.

(above) **Ripening pears** are an unmistakeable autumn sight.

(left) **A trained pear tree** against a wall not only looks good, but also saves space within the garden.

Rootstocks

There are a limited number of rootstock varieties available for pears. The smallest dwarf rootstock is Quince C, which require very fertile soils in order to thrive. Quince A rootstock producs more vigorous trees, and is a good all-rounder, making it the best choice if your soil is not top-quality.

Harvest and storage

Take care when harvesting pears as they bruise easily. Cup the fruit in your hand and gently twist it upwards. If it comes away from the tree easily, it is ready to harvest. If not, leave it for a day or two and try again. Store pears in open trays in a cool place for a few weeks, then bring them indoors where it is warmer to complete ripening. If stored correctly, pears can be kept until late winter, depending on the variety. As with apples, only store undamaged pears.

PLUMS

While plum trees are generally easy to grow, they flower early, which means they may need to be protected from frost damage (*see p.269*).

Do not prune in winter, as cuts can take longer to heal at this time of year, leaving the trees vulnerable to silver leaf disease (*see p.323*). Instead, prune either in early spring or just after harvesting, and generally keep pruning to a minimum, only removing dead or damaged shoots, those that are crossing, and those that particularly spoil the shape of the tree.

The branches of plum trees are quite brittle and break easily, especially when heavily laden with fruit. In a good year it may be necessary to thin fruits to as much as 7.5cm (3in) apart. Alternatively, support heavily laden branches by tying them to a wooden stake hammered into the ground.

If you do not have space to plant several plum trees for cross-pollination, look out for self-fertile varieties. 'Victoria' is a good all-round choice, as it is self-fertile and it tolerates a wide range of climates as long as it is sheltered from frosts.

Heavily laden plum branches should be thinned out or supported with a stake to prevent them from breaking.

Rootstocks

Plum trees can be quite vigorous, so are best grown on dwarfing or semi-dwarfing rootstocks. St. Julien A is a semi-dwarf rootstock that still results in quite large trees, roughly 3.6–5m (12–16ft) in height and spread. For smaller gardens, choose the dwarf rootstock Pixy, which can grow to approximately 2.2–3.6m (7–12ft) in height and spread.

Harvest and storage

Plums don't keep for long, and are best eaten soon after harvesting. They may be stored for a few days in a refrigerator, or otherwise preserved as jam.

CHERRIES

Beloved by many for their snowy spring blossom, cherry trees also provide masses of delightful red fruits in summer – provided that the birds don't get there first. There are two main cherry tree types, based on the fruits they produce: sweet and acid.

Sweet cherries produce fruit on both the current year's growths and on older wood, while acid cherries fruit on the previous year's growths. Because of this, for acid varieties, you need to remove some wood that has borne fruit each year to make way for new wood to fruit the following year.

Like plums, cherries are subject to silver leaf (*see p.323*), so keep pruning to a minimum. Most acid cherries, and some sweet varieties, are self-fertile, so you do not need to plant several varieties in order to produce fruits, as you would with some other fruit trees.

It is rarely necessary to thin cherry fruits, and it is certainly not practical to do so with larger trees.

Rootstocks

Cherry trees can grow quite large, but there are modern rootstocks available that keep them to a more reasonable size, such as the rootstock Gisella 5, which grows to approximately 3–4m (9–12ft) in height.

Harvest and storage

Harvest sweet cherries when they are fully ripe. Eat within a few days, or freeze them. Sour cherries are bitter if eaten raw, but are excellent if made into preserves.

While sweet cherries can be eaten straightaway, sour varieties are best cooked or turned into jam.

PEACHES AND NECTARINES

Little can compare to freshly picked peaches and nectarines. They are closely related; the main difference is that nectarines have a smooth skin, whereas peaches are slightly rougher, but just as tasty. The two fruits are grown in exactly the same way: both are fairly tolerant of soil types but prefer a well-drained soil to crop well.

Peach and nectarine trees are self-fertile. They can flower quite early in the year, when there are few pollinating insects around, so it may be necessary to hand pollinate them. Use a small paint brush to transfer pollen from one flower to another. You may have to do this several times to ensure pollination takes place.

One of the biggest problems when growing these trees is peach leaf curl. This disease, which spreads via water droplets and the wind, causes leaves to become puckered and turn red and purple, with a white powdery spore layer developing on the leaf surface. Keep trees dry with a plastic canopy from mid-winter until late spring, when the disease spreads. Alternatively, if you have a dwarf tree in a container, move it to a sheltered spot during the affected period.

You may need to thin out peach and nectarine trees as they develop, as densely packed fruits may press together and bruise one another as they grow.

Rootstocks

There are dwarf varieties available that make them ideal for growing in pots on the patio in a warm, sheltered spot. The newer dwarf varieties require very little pruning except to thin out crossing shoots or dead growths.

Harvest and storage

Harvest peaches and nectarines when they are ripe or nearly ripe. To test for ripeness, cup a fruit in your hand and gently press close to the stalk with your fingertip. If the flesh is soft, the fruit is ripe and should come away from the tree easily. Place the fruits carefully in a tray so that they do not touch; this helps to avoid damaging the skins. Ripe peaches and nectarines may be eaten right away, while others may need 2–3 days to ripen.

FRUIT BUSHES AND SOFT FRUITS

Most gardeners can find space to grow a few berry or currant bushes. While the likes of blackberry bushes can be quite space-hungry, many modern varieties of soft fruits have been developed to be more compact: some dwarf strawberry and raspberry plants, for instance, are small enough that they can be grown in a container on a balcony. Berry and currant bushes also make an excellent addition to border schemes, with their vibrant fruits acting as a foil to ornamental flowers and foliage.

STRAWBERRIES

Summer wouldn't be summer without strawberries. They do not take up much space and are one of the most rewarding fruits to grow.

Plant strawberries as soon as they are available, from late summer until winter. Plant them in fertile, well-drained soil in full sun, setting the plants 45–60cm (18–24in) apart from one another. The crown of the plant (where the leaves and roots join) should be at soil level to prevent crown rot and other diseases, such as botrytis (*see p.323*). Alternatively, pot strawberry plants up in individual containers with plenty of drainage holes.

If you have a greenhouse or well-lit windowsill, you can bring strawberry plants indoors before flowering. Remember, however, that if you keep strawberries indoors, you will need to hand pollinate the plants with a small paintbrush, as there will be few pollinators around to do the job for you.

Most strawberry varieties are self-fertile. If you want a continuous crop, plant a selection of different varieties with different fruiting seasons, so that you can enjoy fresh berries for as long as possible. After about three years, strawberry plants begin to deteriorate and should be replaced.

(above left) **Mulch around strawberry plants** with a layer of straw in summer to retain moisture in the soil and keep the fruits off the ground.

(below left) **Strawberries are popular** with humans and birds alike. If you must cover plants with netting to protect your crop, make sure that it is tied down securely so that birds do not become trapped inside.

For a long cropping period, choose a mixture of early, mid-season, and late raspberry varieties that each crop for about three weeks.

RASPBERRIES

The taste of a freshly picked raspberry is unrivalled. Raspberry plants are easy to grow and crop heavily, making them an excellent choice if you are new to growing your own fruit. They are also self-fertile, so you will only need one plant in order to produce fruit.

There are two types of raspberries, summer-fruiting and autumn-fruiting. Both types require good, fertile soil and an open, sunny position, and should be kept well-watered during summer dry spells. Their pruning regimes, however, differ depending on when they fruit:

• **Summer-fruiting varieties** fruit on the previous year's growth. Once they have finished cropping, cut down the fruited stems to ground level. It should be quite easy to distinguish these stems from the fresh green growths produced that summer. New growths are not very rigid; tie them into support wires stretched between posts to keep them secure.

• **Autumn-fruiting varieties** fruit on this year's stems. In late winter or early spring, prune the plant back hard by cutting all stems down to ground level. This will encourage plenty of new growths to develop the following spring and summer, ensuring that the next autumn crop will be bountiful.

Newer varieties, such as the summer-flowering 'Glen Ample' and autumn-fruiting 'Joan J.', crop particularly well. By growing a few different varieties, you can ensure a good raspberry harvest from midsummer until the first frosts of autumn. Dwarf raspberry plants are ideal for containers.

BLACKBERRIES

Harvesting and maintaining thorny blackberry bushes used to be quite a painful job, but thankfully, modern varieties have been bred without thorns to avoid this problem.

Blackberry plants are vigorous so need to be spaced around 2.5–3.5m (8–11½ft) apart. Like summer-fruiting raspberries (see p.273), they produce fruit on wood produced the previous year. Cut out the old wood after fruiting and tie in new growths to replace them.

Blackberry plants need to be trained and supported as they grow; otherwise, they become an unruly mess. Fan-training works particularly well for blackberries. This method involves encouraging the stems to grow against a wall or trellis in a "fan" shape, radiating out from the central stem. As new growths appear, loosely tie them together in the centre; then, once the old stems have fruited, cut these out and fan out the new stems to replace them.

REDCURRANTS AND WHITECURRANTS

As well as being delicious to eat, redcurrants and whitecurrants are a beautiful addition to a garden. They are both self-fertile, so you only need one plant in order to produce a crop.

Both redcurrants and whitecurrants are grown in exactly the same way: they like well-drained soil in full sun, although they will tolerate semi-shade. Prune bushes in winter to form a permanent framework of branches, then summer-prune sideshoots in mid- to late summer back to 3–4 buds from the main stems. This encourages "spurs" (short, flower-producing sideshoots) ready for next year's fruits.

Harvest from summer through to early autumn, when the fruits are fully coloured. As the individual currants can be fiddly to pick, they are usually harvested by removing the entire "strig" (cluster of fruits) with scissors. Red-and whitecurrants can be eaten straightaway.

Blackberry bushes are vigorous, space-hungry plants, so make sure to plan for their eventual height and spread.

Redcurrant berries are highly attractive as well as being edible, and make an excellent addition to a border.

While they aren't quite as vibrant, whitecurrants are usually sweeter than redcurrants.

BLACKCURRANTS

Unique among currant plants, the blackcurrant bush is known as a "stool", meaning that its stems develop at or below ground level. Because of this, blackcurrant bushes need to be planted so that about 5cm (2in) of the stems are buried beneath the soil.

After planting, hard prune the stems down to almost ground level. While this means that you won't get any fruit in the first year, cutting the plant down will encourage a good, strong framework of shoots that will then fruit the following year. After this initial prune, cut out 2–3 older shoots to ground level each year to encourage new growth from the base.

Blackcurrants can take up quite a bit of room in a garden. If space is limited, try a compact variety such as 'Ben Sarek', which produces a heavy crop of good-flavoured currants at the end of each summer. This variety can be spaced 1m (3ft) apart, rather than the more usual 1.2–1.5m (4–5ft) required of more vigorous varieties like 'Ben Nevis'.

Rich in vitamin C, blackcurrants can be harvested from mid- to late summer.

TOOLS &
TECHNIQUES
___

Do you really need a garden shed full of expensive, specialized tools and equipment – collinear hoes, daisy grubbers, star-wheeled cultivators, and the rest? The simple answer is "no". Most gardeners – both novice and expert – actually get by with a handful of simple, familiar, well-chosen tools. And less is more in terms of gardening techniques, too. Certainly, when you are starting out, there is knowledge to acquire and experience to accumulate, but gardening is not nearly as difficult as it looks. Much of it is a matter of a little hard work, some regular maintenance, and simple common sense.

Good-quality tools make gardening easier and more enjoyable – and if you look after them they'll probably last you a lifetime.

THE BASIC TOOL KIT

All gardeners have a few favourite tools that they would not be without. When buying gardening tools and equipment, look for good quality products, but don't feel you have to buy the most expensive. In most cases, tools in the mid-price range are fine, but do beware of very cheap equipment that may not last long. A basic tool kit will enable you to do most jobs in the garden. Don't be seduced too early on by the vast range of gadgets on offer. The more you garden, the more you will discover which tools will be helpful and which are gimmicks.

CHOOSING TOOLS

Reputable companies normally offer a few years' guarantee, depending on the tool or equipment. You may even find that a household object or home-made tool does the job just as well. For example, the official tool used to make planting holes for seedlings or young plants is a dibber, but a blunt pencil is a good alternative. Make sure the tools you buy are not only comfortable to grip but also suit your size. If you're not very strong, you may well be better off with a small spade or fork, because larger tools need more leverage to be used effectively. Taller gardeners will almost certainly find that digging is more comfortable when using spades or forks with shafts that are longer than the standard 70–73cm (28–29in). Shafts of up to 1m (3ft) are available.

A pruning saw is useful for cutting small branches; the blade folds away on some types, but one with a fixed blade is best if you have lots of shrubs.

Shears are used for trimming long grass and small hedges, and for cutting back herbaceous perennials. Before buying, ensure they are light enough for you to use.

A rake can be used to level soil or to clear the ground. Clear lawns of leaves and thatch using a spring-tined rake.

A hand fork and trowel are indispensable for weeding, planting young plants and bulbs, and small digging jobs. Hand forks are ideal for heavy clay soils. Small tools are easily mislaid in the garden; look out for those with brightly coloured handles that can be spotted easily, or paint tools with wooden handles to make them more conspicuous.

Secateurs are essential for many pruning jobs; choose a pair with a safety catch and cross-over blades. Try before you buy to ensure they feel right.

A Dutch hoe is invaluable when it comes to clearing a large area of weeds. The long handle helps to avoid backache caused by bending.

Forks and spades have shafts of different lengths and materials: metal shafts are stronger than wood, and some brands may be plastic-covered.

Border spades and forks are a little different as they have slightly smaller heads, making them easy to handle and useful for light work around the lawn and flower beds.

STORING YOUR TOOLS

Keep your tools in a shed, preferably fitted with a lock. Always clean off soil from digging tools before putting them away, and clean and wipe over cutting blades with an oily rag. This will prolong the life of the tools. Invest in hooks, racks, and shelves, or make them yourself out of recycled timber, and get into the habit of hanging up your tools after use and coiling electrical cables neatly. Also clean and stack pots and bundle canes together before storing. By keeping your shed tidy, you will be able to fit more in, find what you need quickly, and avoid unnecessary hazards.

CARING FOR CULTIVATION TOOLS

The majority of digging tools are made from either carbon steel or stainless steel. In order to prevent carbon steel from rusting, it is important to clean and oil your tools regularly. Carbon steel Dutch hoes and spades will slide through soil more easily if you sharpen them occasionally. Some tools have a non-stick coating, but this may eventually wear off. Although stainless steel tools are a bit more expensive, they are worth the money; they are long-lasting, do not rust, and make digging easier, because they cut through the soil cleanly and it does not stick to the blade.

(top) **All metal tools,** including spades, forks, hand trowels, and rakes, need cleaning after use, before you put them away. Brush off any soil, wash, and dry thoroughly.

(middle) **Wooden shafts and handles** of spades and forks need to be checked every so often. Sand off any roughness and wipe over with linseed oil to nourish and preserve the wood.

(bottom) **Carbon steel tools** benefit from a wipe with an oily rag to prevent them from rusting, and blades will cut through the soil more easily if you sharpen them with a file.

CARING FOR
CUTTING TOOLS

Sharp blades ensure cleaner cuts when pruning and trimming. Secateurs need frequent sharpening, as pruning soon blunts the blade. Good quality models are easy to take apart, which makes cleaning and sharpening easier, and allows a damaged blade to be replaced. Knives are often better than secateurs for delicate jobs, and should be cleaned, oiled, and sharpened in the same way. Keep to hand a pocket-sized sharpening steel to sharpen your blade as you work. Use a proprietary tool or a metal file to sharpen shears. Sharpen only the bevelled cutting edge, and tighten the screw holding the blades together.

1 To clean secateur blades, spray on some lubricant or household cleaner to remove any dirt and sap, and then wipe it off with a tissue. For dried-on sap, use a scourer or fine-grade steel wool and metal polish.

2 Smear lubricant or light oil onto the bevelled-edged blade and lightly rub it off with a nylon scourer or fine-grade steel wool. This helps to prevent rusting and keeps the blade smooth between sharpening.

3 Sharpen the blade with an oilstone or a diamond sharpening tool. Hold it at the same angle as the bevelled edge and draw it across the edge. Gently rub the flat side of the blade to remove any burrs, and then oil the blades.

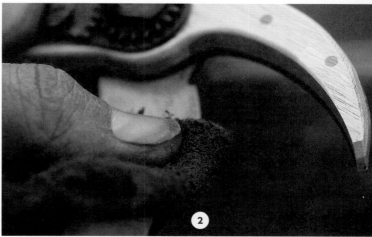

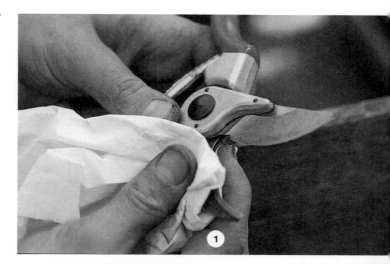

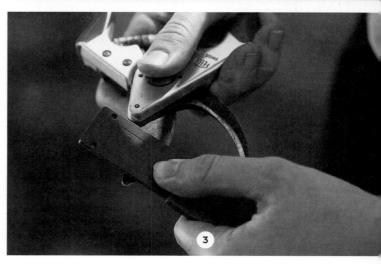

WATERING

Without adequate water, plants will suffer and eventually die. For much of the year, from mid-autumn until spring, rainfall is usually sufficient, but there are times, particularly during late spring and summer, when there is not enough rain. Supplies can run short as rapidly growing plants draw up moisture from the soil, and at such times it is important to know what to water and how.

Create a shallow depression around the plant after planting. Water will then be retained over the area of the roots while it soaks into the soil.

WHAT TO WATER

You do not need to water everything in the garden, even during hot spells. For example, long-established trees, shrubs, roses, climbers, and hardy perennials with extensive root systems can withstand periods of drought. Established lawns, too, can usually do without extra watering; even brown, parched grass will normally green up again when rain returns.

Some plants, however, may need additional watering:
• **New plants** have small, immature root systems that are not yet able to reach water held deep in the soil. Watering after planting – even if the soil does not seem dry – not only keeps the roots moist but also settles the soil around them.
• **Wall-trained plants** can be in a "rain shadow", sheltered from rain by the wall and house eaves. Consequently the soil around their roots is very dry. If you plant about 45cm (18in) away from a wall or fence, the soil will hold more moisture. Mulch plants deeply in spring when the soil is moist to help keep it that way.
• **Newly germinated seeds and seedlings** will die if the roots dry out. Stand pots or trays in water, so that they can take up moisture from below, and then remove them when the top of the soil is damp. This avoids water on the leaves, which increases a seedling's susceptibility to fungal diseases, such as damping off.
• **Conifers** can suffer and may die in hot weather without enough water; if the foliage is badly scorched it may not regrow. Some, such as juniper and pine, are more tolerant of dry soil, and are good choices for warm, sunny gardens or sandy soil.

(above left) **Hanging baskets** tend to dry out quickly in hot and windy weather and therefore need to be watered regularly.

(above right) **Leafy perennial plants** suppress weeds, which take water from garden plants.

• **New lawns** have to be watered until they are well established. Expensive new turf will curl up and die if subjected to drought. Established utility lawns containing ryegrass can do without water. They may turn brown but will green up again after rain.

• **Fruit and vegetable crops** need water most when they are flowering and developing their edible parts. If they run dry at these times, cropping seed in hot, dry conditions is badly affected. Many leafy vegetables wilt or run to seed in hot, dry conditions.

• **Shallow-rooting rhododendrons** fold down their leaves when short of water in hot, dry periods. Flower bud formation is affected, and plants will brown and die if not watered. They are not a good choice for hot spots or sandy, free-draining soils.

• **Plants in containers** need plenty of water because they are growing in a limited volume of compost, which rapidly dries out. In hot, dry, and windy weather check them daily, and twice a day in exceptionally hot weather or if the pots are small.

When planning or adding to plantings, choose plants that match your soil conditions. Drought-resistant plants, such as sedums and lavender, need little or no additional watering; others, such as conifers, will become scorched and remain unsightly for ever if water is in short supply.

PLANTS FOR DRY PLACES

Look out for small, narrow, succulent, or spiky leaves, as these are characteristics of plants that prefer to grow in hot, dry conditions. Silvery or grey surfaces reflect light to keep leaves cool, and a covering of fine hairs minimizes water loss in windy places.

- Achillea
- Armeria
- Artemisia
- Brachyglottis
- Cistus
- Cordyline australis
- Cotton lavender (*Santolina*)
- Eryngium
- Globe thistle (*Echinops, below*)
- Lavender (*Lavandula*)
- Mullein (*Verbascum*)
- Ornamental grasses (many)
- Phlomis
- Phormium
- Rock rose (*Helianthemum*)
- Rosemary (*Salvia rosmarinus*)
- Salvia
- Thyme (*Thymus*)

Auto-irrigation systems can be operated manually or using a timer, providing plants with water on a regular basis. Take care not to flood or deprive pots when regulating the flow.

Lay a seep hose through borders or beds of thirsty plants. This is an efficient watering method; moisture seeps into the soil near the roots where it is needed.

HOW TO WATER

Water early in the morning or in the evening, when cool conditions reduce evaporation, and plants take up less water than at warmer times of the day. Water plants at the base to avoid wetting foliage, which can cause disease. Aim to water plants that are susceptible to slug and snail damage in the morning, as extra moisture in the evening encourages these nocturnal pests.

When watering, a weekly soak is better than a daily dribble. Light watering does not penetrate far, which encourages the roots to grow up to the surface to reach any available moisture – and, of course, they are then even more vulnerable to drought.

WATERING EQUIPMENT

You may not need anything more than a watering can and a hose to water your garden, but do consider other methods of irrigation that may be more efficient and save you time. When using a watering can, fit the rose on the spout, and use a spray adapter on a hose. A fast, solid stream from a spout or hose can easily wash soil from around the plants' roots and may destroy the soil structure. A gush of water is also less likely to penetrate the ground.

• **Sprinklers** should really only be used on new lawns and other large newly planted areas. Avoid using them in windy weather, as the water will be blown off course, or in the heat of the day, when water will evaporate. Rotary types cover circular areas and oscillating designs cover rectangular areas; powerful pulse-jet sprinklers are only suitable for large gardens.

• **Leaky pipes, seep hoses, and drip irrigation systems** are permeable and efficient as very little or no water is lost in run-off or evaporation. Lay the hoses along the base of walls, between rows of vegetables and fruits, and through borders. They can be covered with a loose mulch, such as compost or bark chippings. Most will not work well with the low pressure from water butts.

• **Trickle- or micro-irrigation** systems are useful for pots and in greenhouses. They have a main tube, which you attach to a garden hosepipe, and a number of nozzles or spurs, depending on the area to be watered. The main tube runs from pot to pot, and water trickles out of nozzles or thin spur lines.

• **Automatic timers** set the duration and timing of watering for maximum efficiency, and can be linked to control water to trickle irrigation systems, seep hoses, and lawn sprinklers. Test before you go away.

REDUCING YOUR WATER USAGE

The key to conserving water lies in the soil. Well-structured soil that is high in organic matter retains more moisture than free-draining sandy types; heavy clay holds moisture but plants cannot always extract from it the amounts they need.

Whatever soil you have, you can improve its water-holding capacity by following these simple steps:

• **Dig in bulky organic matter,** such as garden compost or well-rotted manure, on the vegetable plot. Do this every spring on sandy soils and every autumn on heavy soils.

• **Seal in the moisture** with a mulch applied when the soil is moist in spring after rainfall in winter. The best mulch for this is bulky organic matter such as manure or compost; otherwise, use damp newspaper or landscape fabric (geotextile membrane) covered with bark chippings or gravel. Plants in containers can be mulched with decorative materials that may be too costly for large areas (see *pp. 224–227*).

• **Avoid planting and digging** during dry spells. This brings moist soil from the lower depths to the surface, where it will dry out, and sends dry soil from the surface to the lower depths, where the roots need water most.

• **Keep weeds at bay,** because they compete with plants for moisture. Pull them up, or hoe them off while they are young; dig out perennial weeds with the complete root system, or apply a weedkiller (see *p. 295*).

• **Provide shelter** to prevent wind drying out the soil and increasing the rate at which plants lose water through their leaves. Erect a fence or plant a hedge to shield your garden from the prevailing wind. Newly planted trees or shrubs will establish better if protected by a windbreak for their first year; staple windbreak netting or hessian to posts on the plants' windward side.

• **A slow-release watering bag** is useful for newly planted trees or established trees of a certain size. The bag directs water straight to the roots of the tree so there is no wastage.

• **Use water-retaining gel** in containers and hanging baskets, particularly those on a sunny patio (see *p. 229*). This is generally too expensive to add to all beds and borders, but it may be worth considering using it around vulnerable and precious new plants or recent transplants.

• **Choose drought-resistant plants** if you live in a particularly dry area or have a hot spot in your garden. Once established, though, most plants will survive spells of drought.

Water butts can be fixed to the downpipes of your house, as well as to outbuildings, such as sheds and greenhouses. The stored water is suitable for all garden plants except for young plants and seedlings.

MULCHES

A mulch is a layer of organic or inorganic material spread over the soil, where it performs several functions. Any mulch moderates soil temperatures, suppresses weeds, and helps to reduce moisture loss from the soil. Organic mulches, though, are broken down and carried beneath the surface by soil organisms, helping to improve the soil. Inorganic mulches of gravel, pebbles, or sheet materials such as geotextiles, suppress perennial weeds but do not improve the soil structure.

HOW TO MULCH

Mulch between mid- and late spring, when the soil is moist and is warming up. Clear all perennial weeds, then after watering or a spell of steady rain, lay the mulch. Cover the whole surface area of beds and borders, rather than mulching around individual plants. You should only need to top it up once a year in spring. Lay organic mulches 10cm (4in) deep to control weeds and retain moisture, leaving a gap around the base of trees and shrubs so their stems don't rot; you can mulch closer to herbaceous plants, but avoid the crowns of emerging plants. Lay inorganic mulches 2.5–5cm (1–2in) deep.

ORGANIC MULCHES

Organic mulches will decompose, improving the soil structure and adding nutrients.
• **Garden compost** helps prevent weed seeds germinating by excluding light, and breaks down slowly, supplying nutrients gradually.
• **Leafmould** is ideal for woodland gardens or shrub borders and is easy to make.
• **Farmyard manure** must be well rotted. It is a useful source of nutrients, and is good for roses and shrubs.
• **Composted bark,** is the most nutritious of the bark and wood mulches. Use around trees, shrubs, and particularly acid-loving plants.

(above left) **Organic matter,** such as well-rotted compost, makes an excellent mulch for herbaceous perennials. As the growing season progresses, it will rot down and enrich the soil.

(above right) **Spread gravel** around plants that need good drainage, such as sedums. When laid over a weed-suppressing membrane, it reduces maintenance to a minimum.

- **Chipped bark** is low in nutrients and will deplete the soil of nitrogen at first. It is heavy and dense, discouraging weed germination, and lasts for years before it needs topping up.
- **Wood chips** are slow to decompose and initially take nitrogen from the soil. Use for paths or at the back of shrub borders, not around young or herbaceous plants.
- **Composted straw** is low in nutrients and may contain weed seeds, but is fine at the back of a border where it cannot be seen.
- **Cocoa shells** are decorative but costly. They decay rapidly, and need topping up annually. Water to bind them together.
- **Mushroom compost** will supply some nutrients and is slow to decay. Not recommended for use around acid-loving plants like rhododendrons, because it contains chalk.
- **Spent hops** is low in nutrients and lightweight; it may blow around when dry, and rots down quickly. Lay a thick layer and water it.

INORGANIC MULCHES

These are useful for discouraging mosses and preventing any soil from splashing onto flowers and leaves. They provide good surface drainage for plants whose stems and leaves should be kept dry.
- **Gravel** is decorative, and ideal for drought-tolerant plants or around rock and alpine plants.
- **Coarse grit and stone chippings** are ideal for mulching small plants such as alpines or succulents in raised beds or terracotta pots.
- **Cobbles and pebbles** are attractive in most settings, especially around water features.
- **Woven black groundcover** is very useful around newly planted trees and shrubs, helping to retain moisture and suppress weeds.
- **Black plastic sheeting** can be laid around new trees and shrubs with a camouflaging mulch on top, where it will suppress weeds. Biodegradable options are available.

MAKING LEAFMOULD

Autumn leaves are a source of one of the best soil improvers: leafmould. Leaves are broken down by fungi operating in cool temperatures.

Use leaves from deciduous trees and remove any other garden waste. Gather leaves using a rake, and shred the leaves if possible. In a woodland area you may have large quantities, so a vacuum that shreds and compacts leaves can be helpful to speed their breakdown.

Leafmould can be made in a simple wire-mesh enclosure or in a black plastic sack (see *below*), left open at the top for rain to enter. Pack the leaves down and leave them; no turning is needed. Two bins are useful, because leaves may take over a year to break down. One-year-old leafmould should be fine and crumbly, with no discernible leaves (see *bottom left*) – this makes an excellent mulch. After two years, it will be even finer (see *bottom right*), and can be used to improve the soil. Don't add "activators", such as grass, or you will produce compost instead.

FEEDING YOUR PLANTS

Most garden plants survive happily with just an annual application of
organic matter, such as a well-rotted manure or garden compost, which
releases its goodness slowly over the growing season. It also helps retain water
in the soil, which affects the nutrient content too, since nutrients are soluble
and quickly washed away in free-draining soil. That said, there are times when an
additional fertilizer is beneficial, particularly if you are growing fruit or vegetables,
or plants in containers. Fertilizers can also help to kick-start growth after
planting, and reinvigorate plants that have been pruned hard.

Feed shrubs with a general-purpose fertilizer in spring. Sprinkle it around the plant, rake it in, then water.

Yellowing may occur in acid-loving plants on alkaline soil, as they cannot absorb manganese and iron from the soil. Choose plants that will grow in alkaline soil, or apply chelated iron or foliar feeds of manganese.

WHAT TO FEED AND WHEN

Before you buy or use fertilizer, read the instructions on the packet to make sure it is what you want.

• **Vegetables** need a general fertilizer, such as blood, fish, and bone, applied before planting or sowing.

• **Fruiting crops,** such as tomatoes, courgettes, pumpkins, and peppers, appreciate a liquid feed of tomato fertilizer about once every two or three weeks when they are in flower.

• **Permanent plants,** especially fruit, benefit from a general fertilizer, such as blood, fish, and bone, in early spring after weeding and before mulching.

• **Roses** should be fed with a special rose fertilizer in spring straight after pruning.

• **Temporary plantings** in containers and hanging baskets soon use up the limited nutrients in the compost, so use a general-purpose liquid fertilizer every ten days in summer (see *p.229*). For permanent plantings in containers, such as shrubs, simply push controlled-release fertilizer tablets or pellets into the compost in spring (see *p.229*).

• **Lawns** need a fertilizer with a high nitrogen content in spring for lush leaf growth, and one that encourages root development in autumn (see *pp.172–173*).

Regular applications of organic matter help to promote healthy plants; beware of overfeeding, though – this may actually reduce the number of flowers produced by some annuals and perennials.

IMPORTANT PLANT FOODS

The three major plant foods are nitrogen, phosphorus, and potassium, often listed on fertilizer labels as N, P, and K. All plants need these to maintain good health. Other nutrients, such as calcium, magnesium, and sulphur, are required in smaller quantities, while tiny amounts of other nutrients are also vital but rarely deficient in garden soils. One exception is alkaline soil, which is unable to release manganese and iron, and causes yellow leaves on acid-loving plants. A general-purpose fertilizer has equal amounts of the major plant nutrients, together with other nutrients that are not always readily available in the soil.

• **Nitrogen** (N) encourages lush leafy growth, and is the main ingredient in spring lawn fertilizers (see *below*). It is highly soluble and quickly washes out of free-draining soils.
• **Potassium** (K) promotes flowering and fruiting; it is sometimes listed as potash. Fertilizers for tomatoes and summer container plantings are rich in potassium.
• **Phosphorus** (P) promotes healthy root growth, and is essential for young plants whose roots are developing. There is usually plenty found in garden soil in the form of phosphates.

CORRECT FEEDING LEVELS

When feeding, resist the temptation to add a little extra fertilizer for good measure; more plants are lost through overfeeding than underfeeding. Always follow recommended application rates and dilute liquid fertilizer accurately. If in doubt, err on the side of caution, since too much fertilizer promotes lush growth that is more vulnerable to pest and disease attack, and summer bedding tends to flower better if slightly underfed. Overfeeding can cause soluble salts to accumulate and inhibit root activity.

Yellowing between the leaf veins is a common sign of malnutrition. It may indicate a lack of magnesium on acid soils or crops given excess potassium-rich tomato feed; the leaves may also be tinted brown or red. Spray foliage with a solution of Epsom salts every week until symptoms lessen. Potassium deficiency also causes yellowing, purple, or brown tints, and poor flowering, although a lack of flowers may indicate too much nitrogen.

COMPOST

Garden compost is a crumbly, dark, organic material, processed from waste materials from the kitchen and garden. It is made by soil bacteria and other micro-organisms, which break down and rot the raw materials. A good soil improver, compost is easy to add to the soil or to use as a mulch. Grass clippings can be used "raw" as a mulch, but are much more effective once composted.

Wire-mesh enclosures need to be large for good results. Insulate with straw or cardboard between mesh.

Plastic is inexpensive, rot-proof, and the most common material for bins.

BUILDING THE HEAP

Your role in making compost is to provide soil organisms with warmth, moisture, and a good mix of materials. Placing bins on bare soil allows these organisms to get inside. Alternatively, add a spadeful of compost from an old heap, or soil, for every 30cm (12in) of material.

Shredded materials will rot down faster than unshredded ones. You can chop most stems and leaves up with a spade, but it may be worth hiring a shredder in autumn to break down heavier woody material and leaves. It is best to store your compost in a bin; either buy one, or make your own using mesh or wood. A pit is another possibility, but it will be hard to empty and may become waterlogged in the winter. Whichever type of bin or heap you use, it will need a lid to keep out rain.

In theory, you should fill your compost bin with a good blend of materials in as short a period as possible. In practice, it is likely that the bin will take time to fill up. Therefore it probably won't generate enough warmth for thorough composting; weed seeds and roots may survive, as may organisms in diseased material. Large-scale municipal composting reaches temperatures that eliminate these problems, but small volumes of home-made compost cannot match this, so be careful what you add to your heap.

If you cannot achieve the ideal blend of ingredients, you could try using "activators". These nitrogen-rich materials help to break down woody materials, and can be useful when you have too little soft green material. Alternatively, add a thin layer of farmyard or stable manure, mushroom compost, or a sprinkling of nitrogen-rich fertilizer to every 15cm (6in) of woody material. Adding lime is sometimes recommended, but is usually unnecessary, unless you are composting lots of shredded conifer prunings or waste fruit which can be very acidic.

READY TO USE?

Turning the compost can speed up the process. Empty the bin and mix the contents, adding water to dry material before returning it to the bin. To check the progress of your heap, pull back the upper layers to see if the fibrous material is breaking down. If not, it may either be too dry or may need more soft, green material, such as lawn clippings, to add nitrogen.

If you have a small household or modest garden, your compost may not turn out to be the ideal uniformly crumbly, brown material you had hoped for. Instead, it will probably have twiggy and semi-rotted parts mixed in with a dark brown mass that smells like damp woodland. Pick or sieve out the unrotted components, and add them to your next compost batch; they will rot down eventually.

Bad smells indicate compost is too wet – turn it and add fibrous material. A layer of well-rotted compost from another bin or a layer of spent potting compost will also help seal in odours.

COMPOST INGREDIENTS

Getting the right balance is important – ideally 20–50 per cent green, leafy material, and the rest more fibrous, woody material. In practice you will have different materials at different times of year; do what you can with what is available and the materials will balance out over time. The main point is to try to prevent the bin being dominated by one ingredient.

Put it in
Green materials such as:
- **Vegetable kitchen waste**
- **Weeds** that haven't gone to seed
- **Grass clippings**

These provide nitrogen and other nutrients for the micro-organisms. They are wet and soft, so must be mixed with fibrous material.

Woody materials such as:
- **Shredded paper** or cardboard
- **Eggshells** and carrot peelings
- **Spent bedding plants**
- **Fallen leaves**
- **Decaying stems** of perennials
- **Twiggy prunings**

These provide tougher, carbon-rich material with less nitrogen. If you don't have time or space for leafmould, fallen leaves can be included.

Leave it out
- **Material that is diseased,** damaged, or contaminated with weedkiller
- **Weeds carrying seeds** or with roots that might survive composting, such as dandelions or bindweed
- **Cat and dog droppings,** which may harbour harmful organisms
- **Kitchen waste** containing animal materials, such as scraps of meat, which can attract rats.

DIGGING

Wild plants get along perfectly well without any digging, but in gardens some cultivation is useful. It loosens packed down or "compacted" soil, improving aeration and drainage, and burying or loosening weeds, making them easier to pull out. When cultivating the soil, you can also incorporate any organic matter, fertilizer, or lime that you apply, far more quickly than rain or worms can. An inspection of your soil should give you an idea of what is needed (see *pp.18–19*).

HOW AND WHEN TO DIG

Digging is slow and heavy work, especially on clay soils, so you may want to do it in stages. Sometimes it is enough to loosen the soil with a fork or large hoe to mix in fertilizers or lime – a good technique for sticky clay soils. Forks and hoes are also useful for working around existing plants and teasing out weed roots. Soil may only need digging once every two or three years; you can often avoid it altogether by keeping off the soil, and only working it in dry weather. If soil sticks to your boots when you walk on it, it is too wet to work. The ideal time to dig clay and loam is autumn; winter frosts will help to break down any clods on the surface. Sandy soils, however, are best dug in spring.

DIGGING IN ORGANIC MATTER

Adding organic matter ensures that soil is fertile – full of living organisms and nutrients that growing plants need, particularly nitrogen. Fertilizers, in contrast, aim to feed the plants, not the soil.

Organic matter will also help to improve soil structure. Well-structured soil retains nutrients and moisture well, allows excess water to drain away, and traps air needed for roots to breathe. The humus in organic matter bonds fine clay particles into crumbs, creating larger spaces between them and improving drainage. Humus also improves sandy soils, increasing both their water- and nutrient-holding capacity.

(top) **Lay paths** or work from planks on the soil to preserve its structure and prevent compaction.

(bottom) **Raised beds** help to reduce labour. Keep them narrow to tend them from the path, without treading on the soil.

(opposite) **Incorporating organic matter** (such as well-rotted manure) into the soil as you dig will add nutrients and improve the soil structure.

HOW TO SINGLE DIG

Single digging is so-called because it involves turning over the soil to the depth of a single spade blade. It is done to loosen soil, bury weeds or plant debris, and mix in organic matter, fertilizer, or lime.

Start by dividing your plot into two adjacent rectangular strips, each about the width of a spade and as long as the plot. The idea is to remove soil from the first strip, and replace it with soil from the second, and then place the excavated soil from the first strip into the second trench. In effect, this is simply swapping the soil between the two trenches. Repeat this technique across your plot to dig over all the soil.

1 Start by digging out the soil from the first strip. Work from the front to the back of the strip, and create a trench one spade or "spit" deep and one spade wide. Place the excavated soil on the surface to one side of the trench.

2 Add organic matter, such as well-rotted manure or garden compost, to the bottom of the trench. Then dig out the second strip as in step 1, but place the soil into the first trench, so that it covers the organic matter that has already been added.

3 Add some more organic matter to the second trench, using a fork to break it up if necessary. Fill in this second trench using the soil dug out from the first. You can dig larger plots in the same way, dividing them up into several strips.

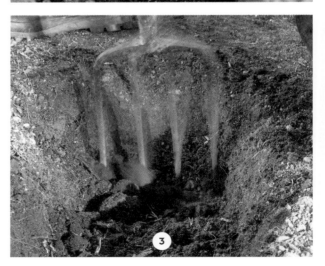

WEEDING

There are several good reasons to keep your garden clear of weeds. They not only compete with flowers for light, nutrients, and water, but some, such as chickweed, also harbour diseases like cucumber mosaic virus, which may spread to surrounding plants. Weeds also look unsightly, and some of the more pernicious types, such as ground elder and bindweed, can swamp a bed or border within one season if left unchecked.

• **Start weeding early,** before weeds have a chance to flower and spread their seed.

• **Remove the whole weed,** roots, stems and all. Some stubborn perennial weeds, such as couch grass and bindweed, spread by underground roots and stems. Others have a deep taproot, like dandelions. Try to dig out every bit of these weeds, as they can regrow and reproduce from even a tiny piece of root.

• **Use a systemic weedkiller** for perennial weeds; it will spread from the leaves to kill the roots, breaking down in the soil without harming other plants. It is available as a liquid, or a paint-on gel for weeds in borders. Cover surrounding plants when applying to ensure it does not touch them.

• **Avoid harming other plants** when removing weeds; digging the soil between plants can damage roots and will bring more weed seeds to the surface to germinate. When weeds are close to other plants, hoe as shallowly as possible, and remove weeds carefully by hand to minimize soil disturbance.

• **Lay a weed-suppressing mulch** to exclude light if your soil is full of weeds. Weed seedlings need light to grow, so this should stop any further ones appearing.

(top) **Slice the roots** just under the surface when hoeing off annual weeds. Hoe on a dry day, leaving the weeds to wither, then compost them.

(bottom) **Stepping stones** in beds and borders allow you to keep off the soil while you weed, reducing soil compaction and the need to dig.

ECO-GARDENING

The pros and cons of organic gardening are hotly debated. In truth,
most of us would prefer not to use chemicals if we could avoid the need
for them, and we would be happier to rely on natural, organic products to
ensure our plants remain healthy and well-fed. There may be times when it is
hard to resist synthetic solutions, but there are plenty of ways in which you
can make your garden as green and eco-friendly as possible.

(above) **An environmental garden** made entirely from recycled products includes reclaimed stone and timber paths, a living turf roof, and even recycled topsoil.

(left) **Attract bees** to your garden by growing pollen-rich flowers such as roses and geraniums. Even weeds such as this sow thistle are a rich source of pollen.

ORGANIC OPTIONS

Incorporating organic matter into your soil should supplement it with all the nutrients that plants need for healthy growth. Both manure and compost are particularly good for replacing nitrogen, which is soluble and easily washed out of soil by rain, or lost through the action of certain types of bacteria. Moreover, they release it gradually as they decay, thus feeding plants throughout the growing season.

Biological controls

Some of the creatures you might initially think of as pests are in fact the very opposite: they're natural predators. And they can be employed to tackle the bugs that really threaten your plants.

A combination of mixed planting and leaving a few areas of your garden less than immaculately tidy will help attract bees, ladybirds, hoverflies, lacewings, and other beneficial insects. You could also dig a pond to provide frogs and toads with a home. You might even fancy introducing some specialist contract killers yourself – slug-destroying nematodes, for example.

Chemical-free gardening

There are alternatives to synthetic pesticides and weedkillers. Derris, pyrethrum, and insecticidal soaps are effective against certain insects and are all derived from plant extracts. Weeds that can't be dug up can usually be killed off by covering the ground with plastic or by using a flame weeder.

WISE WATERING

With pressure on water supplies growing and summers arguably becoming increasingly dry, water conservation is important. A water butt can collect rainwater from down pipes or from shed and greenhouse roofs. In larger gardens you can link butts together with an overflow pipe. Make sure the butt stands on a hard, stable surface, and is high enough for you to get a can under the tap. Cover the top with a lid to keep out debris. Conservation issues aside, rainwater is best for plants; in some areas tap water contains high levels of lime and is described as "hard"; this is unsuitable for acid-loving plants such as rhododendrons. Grey water, or waste household water, can also be used in the garden, especially when a hosepipe ban is in force. Do not use water that has passed through a water-softening unit or dishwasher, as it contains chemicals that damage plants. You can use washing-up water that is not fatty or full of detergent, bath water with no bath oil and not too much bubble bath or shampoo in it, and water from rinsing clothes. Let the water cool, and then use it as soon as possible to prevent bacteria from building up. Water onto the soil, not over the foliage. Do not add grey water to a water butt or use it in irrigation systems, and avoid using it on acid-loving plants, edible crops, seedlings or young plants, and lawns. If you want to recycle grey water regularly, you can buy adapters to take water from waste water pipes.

Dry gardens, in which water is scarce, need plants that thrive naturally in arid conditions. Gravel is a better choice than a lawn that demands regular watering.

PROPAGATING PLANTS

Most plants reproduce by seed, but this can be hampered by several factors. For example, adverse weather conditions can reduce insect populations and pollination. To counter such problems, plants have evolved back-up methods. Perennials often form large clumps made up of many small plantlets that can survive on their own. Other plants can regenerate from a section of root or stem separated from the main plant, and the flexible stems of climbers and some shrubs take root if they rest on the soil. These traits are exploited by gardeners to make new plants.

Plant propagation in the wild requires no equipment at all: just turn an oriental poppy seedhead upside down and look at how many seeds it contains, each one holding out for the right conditions for germination.

ESSENTIAL EQUIPMENT

A few basic items are essential for successful propagation. You will need:
• **A pencil or waterproof pen** and labels for identifying plants
• **Secateurs and a sharp knife** for taking cuttings
• **Plastic bags** and elastic bands to secure them over pots
• **A small sieve** for dusting compost over seeds
• **A block or piece of board** for firming compost before sowing
• **A dibber or pencil** to make planting holes and lift seedlings.

TRAYS, MODULES, AND POTS

Seed trays are useful for raising large quantities of seeds. The usual size is about 35 x 23cm (14 x 9in); half-sized trays are also available. Module or multi-celled trays have a number of internal divisions, each treated as a separate small pot. These eliminate the need to prick out young seedlings (see p.303) and are useful for minimizing root disturbance or growing several different plant varieties together. Root-trainers are deep modules with ridged sides to help long roots grow straight.

For single or small numbers of seeds or cuttings, 8-9cm (3-3½in) pots are the most suitable size; for larger quantities 13-15cm (5-6in) pots are more economical. Plastic pots are easier to use than clay ones, and retain moisture for longer; either type must have drainage holes at the bottom. Biodegradable pots eliminate root disturbance when seedlings are transplanted to larger pots or outside into the ground. As your seedlings grow, they will have to be moved outside to harden off (see *p.303*).

KEEPING SEEDLINGS COVERED

Most seeds and cuttings need protection from drying out. Small sheets of glass or clear plastic lids are ideal for seed trays. Enclose individual pots in clear polythene bags held clear of leaves by canes or hoops of wire. Propagators consist of a tray and clear plastic cover, and are available in various sizes. They keep seeds and cuttings warm and the air inside humid. Heated propagators, especially those with an adjustable thermostat, can maintain higher temperatures to speed up rooting.

For propagation outdoors, glass or plastic cloches offer a little protection from low temperatures and bad weather, and are most effective if you can regulate their ventilation and open them easily for watering.

CHOOSING SOIL AND COMPOST

You will need a suitable compost or rooting mixture in which to sow seeds or grow plants. Some plant types need special composts: alpines, for example, grow best in very sandy or gritty conditions, while acid-loving plants such as heathers and pieris must have a lime-free ("ericaceous") compost.

• **Seed or cuttings composts,** such as a specialized seed potting media, are ideal; potting compost is too rich for seedlings. Soil- or loam-based types dry out slowly. Soilless mixes are light and clean, but can be hard to re-moisten once dry.

• **Horticultural grit** on top of compost helps water drain away from stems; in the base of a pot it prevents waterlogging. When sowing water-sensitive plants or rooting cuttings in a cool season, add coarse sand for drainage.

• **Vermiculite** is a sterile, lightweight material that holds moisture and improves aeration. Choose a grade that is intended for propagation. Perlite is similar – both are added to compost or used on their own for rooting cuttings.

HOW PLANTS REPRODUCE

The creation of seeds is dependent on the process of pollination, in which pollen grains produced by the male stamen of a flower are transferred to the female stigma. Pollen is distributed by insects such as bees or may be carried by the wind, as with dandelions (see *below*). Some plants can pollinate themselves ("self-pollination") but others need pollen from another plant of the same species ("cross-pollination"). A small number of plants, such as citrus, are apomictic, meaning that they set seed without pollination.

If pollination is successful and the pollen grains germinate, seeds form inside the ovaries at the flower base. The seeds ripen, disperse, and under the right circumstances they themselves will germinate, growing into new plants.

GROWING PLANTS FROM SEED

Growing from seed is the most common and often the easiest propagation method. All the seed needs to grow is warmth, moisture, light, and air. If you provide these at the right time, you should be successful. If you want to avoid pricking out seedlings later, sow in modules, a few seeds per cell. After germination, pinch out all but the strongest seedlings.

Buying and saving seeds

Most seeds are sold in a plastic or foil sachet that preserves them until opened. Seed in paper packets ages more rapidly. Keep the packet, as it will give growing recommendations and a "use by" date, after which unused seed deteriorates.

Home-grown seed is usually worth saving, but choose healthy plants with the best seedheads. Seeds must be mature, but not so ripe that they are already being shed. Enclose dry heads and capsules in paper bags before cutting them off, then label and store the bags. Spread damp seeds out to dry in a warm, airy place. Dry seed pods and store well, or extract and dry the seeds

Storing and treating seeds

Keep bought seed in its packet (re-sealed if it has been opened), and saved seeds in envelopes or paper bags marked with the date of harvest. Store in a dark, dry, well-ventilated place, or in tins with one or two sachets of silica gel to absorb moisture. Aim for conditions that are frost-free but below 5°C (41°F).

A few seeds need special treatment to speed germination; this will be detailed on the packet. For example, hard seed coats that are slow to absorb moisture can be nicked with a knife (chipped) or rubbed with sandpaper (scarified), or the seeds can be soaked in warm water overnight. Other seeds may need to be chilled in the refrigerator to simulate winter conditions.

SOWING UNDER COVER

Tender plants and half-hardy annuals and perennials are started indoors since they need artificial warmth or frost-free conditions. The seeds of most hardy plants can be sown indoors too, if you have nowhere outdoors to sow them or you want closer control of their germination and early growth. Remember that seeds sown indoors must be checked regularly to ensure that they are warm and moist. Sow at the recommended depth, as either too much or too little light can prevent the seeds from germinating, and sow sparingly both to reduce the risk of overcrowding and disease, and to make pricking out easier.

1 Fill the container to the brim with compost. Tap it lightly on the work surface to settle the contents, and level off any excess. Firm to 1cm (½in) below the rim to eliminate air pockets. Use a presser of wood or plywood or the base of another pot.

2 Pour some of the seeds into the palm of one hand, and use finger and thumb of the other to sprinkle them over the surface. Never sow direct from the packet, as it is hard to shake seed out evenly, and seedlings are likely to end up in overcrowded patches.

3 Sieve a shallow layer of compost over the seeds if they need darkness for germination. Fine seeds needing light, such as begonias, can be pressed gently into the surface or covered with a thin layer of vermiculite. Label with the variety and date of sowing.

4 Water from below by standing the pot in shallow water until moist patches appear at the surface. Drain and cover with a clear polythene bag, or enclose in a propagator. Keep at the recommended temperature and out of direct sun; cold nights or very hot days can cause failure.

LOOKING AFTER SEEDLINGS

The time it takes for seeds to germinate and produce a visible shoot above the soil varies greatly, ranging from a day for plants like cress and radishes, to a year or more for some shrubs and trees. Germination starts with the appearance of a root, followed by a shoot. Note that the first pair of leaves to develop may look different from those that follow. They are called the "seed leaves".

If you are germinating seeds in darkness, check them regularly and move into good light once the seedlings emerge to encourage sturdy and healthy growth. They will grow tall, thin, and pale if left in the dark, but keep them out of bright, direct sunlight, which can scorch tender leaves. Keep seedlings warm, although they usually tolerate lower temperatures than those for germination.

Do not let the seedlings dry out at any stage, but avoid overwatering, which can drown the roots. Use a copper fungicide to reduce the risk of damping off disease. Water small seedlings from below as for seeds, but more robust kinds, such as lupins and peonies, can be watered gently from above. Use a watering can with a very fine rose, and start and finish with the can to one side of the container to avoid letting heavy drips fall on the plants. Allow containers to drain thoroughly after watering, and never keep them permanently in trays or saucers of water.

Getting ready for the garden

When you have pricked out your seedlings, grow them on at the temperature recommended on the seed packet. Check them daily and water whenever they look dry. Plants in trays can usually remain there until they are planted out, but some vigorous seedlings in pots may need to be moved again into containers 3–5cm (1–2in) wider than their root systems. If planting out is delayed and the foliage starts to turn yellow, feed with a balanced liquid fertilizer.

Plants that have been raised indoors make soft growth that is vulnerable to wind and cold. They must gradually be acclimatized to conditions outdoors – this is known as "hardening off". Several weeks before

Use plug plants if you don't have time to raise plants from seed. Water them when they arrive, and use the blunt end of a pencil to push the plug plants out of their modules.

Grow plug plants on in module trays filled with compost, making holes using a dibber or pencil; keep the plants in a light, airy place. Water from below by placing the module in a tray of water.

planting out, place plants in a cool spot indoors. A week or so later, stand them outside for a few hours each day somewhere that is both sheltered and shady, covered with a double layer of horticultural fleece. It is important to bring them in at night. Leave them out for a little longer each day and then remove one layer of fleece. About a week before you plant them out, remove the fleece during the day, but leave them out at night still covered by the fleece.

If you have an unheated cold frame, place the containers of plants in this instead of using fleece, keeping the lid shut for a few days. Gradually increase the ventilation, opening the lid just a crack to start with but eventually leaving it fully open, at first during the day and then at night too.

Planting out

Plant out as soon as the young plants are hardened off and the weather is suitable: mild, damp conditions are best. Wait until all threat of frost is past before risking half-hardy plants outdoors. Prepare the site thoroughly in advance, and water the plants well and leave to drain before planting.

Tap each side of a tray sharply on the ground to loosen the compost, then slide it out in one block and separate the plants carefully with your fingers or a trowel. Release plants from modules by pushing on the base of the cells; some modules are torn apart. Check the seed packet for planting distances. Make a small hole for each plant, carefully place it in position, draw the soil over the roots, gently firm it down by hand, and water it in.

PRICKING OUT SEEDLINGS

Seedlings in trays become overcrowded quickly, leading to weak growth. To avoid this, "prick out" seedlings when they are large enough to handle, discarding the weakest. Always hold them by a leaf to avoid damaging the fragile stem.

1 Separate the seedlings using a dibber or pencil, and remove each one gently by its leaves. Fill pots, modules, or trays with compost, and make a hole for each seedling.

2 Transplant a seedling into each hole, at the same depth as before or slightly lower. Firm around each one gently, then water well.

DIVIDING PERENNIALS

Select a healthy plant and water it a few hours before lifting to moisten the roots. Incorporate well-rotted manure or compost into the soil in the new location, along with some all-purpose fertilizer if needed. If replanting is delayed, dip the divisions in water and keep in a plastic bag in a cool place so they do not dry out.

1 Cut down old stems so that you can see the crown. Dig around the clump and lever it out with a fork. Shake off surplus soil, and wash out any weed roots. Try teasing apart the clump by hand. If this is too hard, insert two forks back to back and work the handles together to split the clump. Repeat until you have enough good-sized segments.

2 Replant segments in fresh, well-prepared ground before the roots have a chance to dry out. Check the plants are at the same depth as before, firm in and level the soil, and water well.

DIVIDING PLANTS

Division is the easiest method of making new plants and is the usual way of increasing most herbaceous perennials. Many hardy plants mature and fatten into bushy clumps which can be divided to create new rooted plants. Unlike growing from seed, division uses an existing plant to produce more plants that are exactly like the original. It is also a reliable way to rejuvenate perennials that are old and no longer flowering well.

Choosing plants to divide

The plant you select to divide must be healthy in every respect. Whereas few ailments are passed on through seeds, a diseased parent plant used either for division or for taking cuttings (see pp.308–311) will usually result in diseased offspring.

When buying new herbaceous perennials, look out for large plants: these are an economical option if you can divide them into two or three portions before planting. In choosing a suitable clump of plants for division, check that the roots are healthy. They should be flexible and undamaged; most healthy roots also have numerous short, pale, hair-like side roots growing along them. Discard plant segments that look dry or withered, with dark roots that crumble or break easily, since these may be symptoms of disease. Discard the tired, old parts of the plant, which are usually in the centre of the clump, and select the young, vigorous sections towards the edges for replanting.

When to divide

Most perennials can be divided at any time while they are dormant and unlikely to suffer any check to growth, but different times of year are better for different divisions.
• **Early spring** is the best time to divide most perennials, especially the fleshy-rooted types. The worst weather is usually over and growth will resume quickly. Some plants will already have buds, which can help you to distinguish the strongest parts for replanting.
• **Once flowers start to die down** in early-flowering plants – such as doronicum, pulmonarias, and *Primula denticulata* – you can start to divide them. This is also the best time to split clumps of bearded irises.
• **Between late spring and early summer,** divide marginal and aquatic pond plants – this is when their growth revives.

Clump-forming herbaceous perennials can often be divided by hand after they have been dug up. The smaller portions can be replanted to create new rooted plants.

DIVIDING BEARDED IRISES

Bearded irises produce thick horizontal stems ("rhizomes") at soil level. Lift these with a fork, taking care not to damage the fleshy rhizome. Shake off any soil and weed roots, and divide them with a knife into short rooted segments, each with at least one growth bud.

1 Split the rhizomes into sections, so that each has healthy shoots and good roots. Trim off the older sections furthest from the leaves with a knife, and then dust cut surfaces with fungicide.

2 Trim long roots back by one third, and cut leaves to about 15cm (6in). This will reduce rocking by the wind, keeping the replanted section stable until it develops strong, anchoring roots. Replant the sections at the same depth as before in well-prepared ground, spaced about 12cm (5in) apart. Firm in, level, and water well.

• **During late summer,** some perennials, such as hostas, often make new roots, and it is a good time to divide them. Most of the growth is over and divisions will quickly settle in. To be on the safe side, keep late divisions in a cold frame or nursery bed to protect them from harsh winter weather, and move them to their final positions in spring.

• **Avoid winter,** as the weather can be wet or very cold, with a risk that divisions will rot.

Dividing techniques

Herbaceous perennials may develop solid, woody crowns, clusters of smaller crowns, or simply a mass of fibrous roots. Chop through solid crowns with a spade, or cut them with a knife. Ensure each segment has plenty of healthy roots and several strong growth buds. Divide looser clumps with garden forks, or hand forks if they are small. You may be able to simply tease them into smaller portions with your fingers. Some plants, such as alpines, produce "offsets", complete young plants alongside the parent. Loosen them with a hand fork, separate them from the parent plant with a sharp knife or secateurs, and replant.

Divisions that fit comfortably in one hand are the best size for replanting, but you can use larger ones for rapid establishment. Grow on smaller fragments with one or two buds in a nursery bed without any competition for a season before planting them out.

Tip layering should take place in summer. Select a healthy young stem and bury the shoot tip in a hole 7–10cm (3–4in) deep. A new shoot will appear from the stem tip after a few weeks; sever it from the parent plant. Allow the new plant to grow on, and transplant the following spring.

SELF-LAYERING CLIMBERS

Some climbers, such as honeysuckle and ivy, naturally take root where their stems bend down and touch the soil. You can sever rooted sections in the autumn, or in spring if they have not formed a strong enough root system in autumn. Choose a rooted section with new leaves beyond the roots, and lift it out of the soil with a hand fork. Then trim off all other parts of the stem, except for those with new roots and leaves. Remove the lower leaves on the rooted section of stem, and then plant it up in a pot or in its final position.

LAYERING PLANTS

Many shrubs and climbers will naturally develop new roots on low branches if these come into contact with the ground or are buried for any length of time. You can use this phenomenon to produce new plants. Known as layering, it is a particularly attractive method, because the stem used is left attached to the parent plant until it has rooted, so nothing is lost if the attempt fails. This is a useful way to fill a low gap in a mature hedge.

How layering works

You can simply bury a stem to exclude light and it might start to root, but the chances of success are often increased by cutting or twisting it to damage some of the tissues. This interrupts the normal flow of hormones and other sap-borne chemicals, encouraging them to become concentrated at the injury site and stimulate root formation. The section of stem beyond the wound will become slightly water-stressed – this also promotes root growth as the plant struggles to survive. Applying a rooting hormone powder or solution to the wound accelerates the process further. Rooting takes a variable amount of time, depending on the species and when you layer the stems: do it in early spring, and most layers will have rooted by the autumn.

Types of layering

• **Simple layering** involves burying a single shoot until it takes root, then severing it in order for a new plant to establish itself (see *opposite*).

• **Serpentine layering** involves pegging down the stems of climbers like clematis, honeysuckle, vines, and wisteria a few times between leaves to produce new plants. Make several wounds along a young stem, each behind a leaf joint or bud. Peg down the wounded sections, leaving the stem between exposed to the light. Roots will form on the buried wounds, while buds on the exposed sections will develop into new shoots. When well-rooted, separate and transplant sections.

• **Tip layering** is used for blackberries, brambles, and other *Rubus* that naturally produce roots from their stem tips (see *above*). It should be carried out during the summer.

• **Mound layering** is also known as "stooling". In spring, soil is mounded up around the crown of low-growing shrubs such as heathers, rosemary, thyme, and lavender so that they are half-covered. The stems will start to develop roots and can be uncovered, cut off, and transplanted in late summer or autumn.

SIMPLE LAYERING

This easy method can produce new plants within a few months. Once you have layered a shoot, it will need very little attention, because it is still receiving most of its water and nutrients from the parent plant while it forms new roots. Only once you have severed that connection do you need to take extra care that the new plant does not dry out.

1 Select a vigorous, healthy shoot that can be bent to touch the ground with little effort. Trim off any sideshoots and make a shallow, slanting cut on the underside of the shoot about 30cm (12in) from the tip. Applying a hormone rooting compound on the cut should increase the chances of success.

2 Carefully bend the stem to the soil. Weight or peg it down in order to keep the wound in contact with the soil, and tie its tip to an upright cane. If the soil is poor, bury the wounded area in a shallow hole filled with moist compost. Firm in gently with your fingers and water well.

3 When the layered shoot is well rooted and the tip has started to grow new leaves, you can transplant it into a pot or its new position. Sever it from the parent plant close to the new roots.

TAKING HARDWOOD CUTTINGS

Use this method for deciduous trees, shrubs, roses, and climbers. Choose strong, straight, well-budded stems of the current year's growth once the leaves have fallen. For trees, leave a single bud above the ground; for multi-stemmed bushes, allow 2.5–5cm (1–2in) of stem above ground so that several buds develop into shoots.

1 Make a V-shaped trench, 20cm (8in) deep, adding horticultural sand if drainage is poor. Remove a long stem, then cut it into lengths of about 23cm (9in), pruning the bottom below a bud and the top just above another bud, with a sloping cut to distinguish it. Remove any leaves and sideshoots.

2 Insert the cuttings upright 10–15cm (4–6in) apart along the trench, with enough buds above the surface to form a few shoots. Backfill around them with soil and gently firm, then label and water well. Root less hardy plants, such as perovskia, hibiscus, and cistus, in pots in an unheated greenhouse until spring. Trim cuttings to 8–10cm (3–4in), insert with the top bud just above the surface, and keep moist. Plant out next autumn.

GROWING PLANTS FROM CUTTINGS

Taking cuttings is probably the most popular propagation technique after sowing seed. As the name implies, cuttings are portions of plant stem or root (occasionally leaves or a bud) cut from a strong, healthy parent plant and encouraged to develop their own roots. This technique works for the majority of plants and produces exact replicas of the donor. Unlike divisions or layers, cuttings are dependent on your care and a supportive environment until they are self-sufficient. Some are quicker to root than others, and different types will be more successful if taken at a particular time of year.

Types of cutting
Cuttings can be taken from both stems and roots:
• **Stem or shoot cuttings** are distinguished by the age and maturity of the shoots used.
• **Root cuttings** are sections of root, normally taken from the parent plant when it is dormant in winter. The downside is that the parent plant will need to be dug up.

Stem cuttings also differ according to when they are taken:
• **Hardwood cuttings** are taken from mid-autumn onwards, when shoots are woody and the leaves have fallen. Although the slowest of the stem cuttings to root, most are simple and can be grown outdoors.
• **Softwood cuttings** root quickly and are best taken in spring and summer, using the soft, new tips of the current year's young shoots before they start to become woody.
• **Semi-ripe cuttings** are taken from early summer to early autumn, when the bases of the shoots are firm, or "ripe", but the tips are still soft and green. Because the cuttings are firmer, they are often easier to deal with than soft cuttings.

Essential equipment
You will need both secateurs and a knife; clean them before taking each new batch of cuttings and keep them sharp. Use fresh compost and new or sterilized containers. Cuttings must be spaced out when planted, so that their leaves do not touch. Small pots, about 9cm (3½in) diameter, are ideal for single cuttings; a 13–15cm (5–6in) pot will accommodate several cuttings. Larger quantities can be rooted in a module or cell tray at least 5cm (2in) deep. Use a cuttings compost, which is blended to be

About six months after planting a root cutting, leaf growth may appear before plants have had a chance to make good root systems. Wait until you see roots near the holes at the bottom of the pot before transplanting.

free-draining, or mix your own from equal parts soilless seed compost or sieved leafmould and sharp horticultural sand or grit.

Natural warmth will often be enough to root cuttings in summer if pots are enclosed in plastic bags or trays are covered with lids. At cooler times, or if you produce lots of plants from cuttings, a propagating case on a heated tray or a heated propagator fitted with a thermostat will remove much uncertainty. Hormone rooting preparations can speed up root initiation; buy some each season and store in a refrigerator.

Taking root cuttings

This is a very easy and reliable way to produce good results. It is the best way to multiply perennial border phloxes, because it does not transmit eelworm, a pest that infests their top-growth. Root cuttings of variegated plants will produce all-green plants.

Cuttings are normally taken during the plant's dormant period in midwinter, to minimize disturbance and injury. Always choose a healthy plant. Lift small plants out of the ground completely with a fork and replant immediately after taking the cuttings. With larger plants, scrape away enough soil to expose the roots, and then replace and firm the soil at once to avoid destabilizing the plant. Never remove more than a few roots from each plant.

Most plants propagated this way have thick roots, which are planted upright. Plants such as phlox and *Primula* denticulata have thin roots, which are rooted horizontally in trays.

TAKING SOFTWOOD CUTTINGS

Softwood cuttings are taken from soft, new growth at the tips of non-flowering shoots, produced in spring and early summer. Most root in six to eight weeks. Softwood shoots wilt quickly; take cuttings early in the day before the sun gets hot. Delphiniums and lupins can also be increased from basal shoots, which sprout from the plant's crown in early spring. Cut them when they are 5–8cm (2–3in) long, with a portion of the crown at their base, and root them in the same way.

1 Cut sections 8–10cm (3–4in) long from the tips of healthy, young, non-flowering stems, using secateurs or a knife. Place in a closed plastic bag out of the sun as you work. Trim the cutting just below a leaf joint. Remove the leaves from the lower part of the cutting so that only 2 or 3 remain at the top.

2 Dip the end of the stem in hormone rooting material. Push the lower half into compost, around the edge of the pot when there are several cuttings. Gently firm in, water thoroughly with tap water, and label. Enclose in a plastic bag or propagator. Most root best at 15–21°C (59–70°F).

TAKING SEMI-RIPE CUTTINGS

This is often used to propagate shrubs that do not grow well from hardwood cuttings, especially broad-leaved evergreens such as aucuba and mahonia. It is also used for climbers, conifers, and tender perennials. Between midsummer and mid-autumn, gather strong, healthy sideshoots, about 15cm (6in) long. Some plants root more successfully from sideshoots pulled (rather than cut) off with a short "heel" of bark from the main stem. Long year-old stems can make several cuttings, each trimmed below a node at the base and just above one at the top. Cuttings are usually rooted indoors; most semi-ripe cuttings take six to ten weeks to produce roots; those taken in late summer may even take until the following spring.

1 Choose a current season's shoot that is slightly resistant to bending at its base but still soft and green at the tip, and cut it just below a leaf joint.

2 Trim off the soft top portion of the cutting just above a leaf using a knife, then discard it. This reduces moisture loss through the soft leaves at the tip.

3 Make a shallow cut 1–2cm (½in) long on one side of the stem base to encourage rooting. Remove the leaves and sideshoots from the lower half of the stem.

4 Dip the bottom 5mm (¼in) of the cutting in hormone rooting material, and tap off any surplus. Dust the entire wounded area with the powder.

5 Insert the cutting up to its lowest leaves in cuttings compost, or 8cm (3in) apart in the soil in a cold frame. Firm in place, label, and water with tap water.

6 Place the pot in a lidded propagator, or cover with a plastic bag. Hold the bag clear from the leaves by inserting sticks into the sides of the pot.

GARDEN DOCTOR

———

All gardens, however carefully nurtured, are likely to harbour some pests and diseases. Usually, plants' own natural defences are able to fend them off. Most can tolerate some damage, and will survive a minor onslaught. However, as any doctor will tell you, prevention is always the best cure. Plants growing in situations to which they are not suited are less able to resist attack, as are those starved of water, light, and nutrients or exposed to extremes of heat, cold, and wind. So, site your plants in the appropriate place, and keep them well watered and cared for. Vigilance is important: take a few minutes to stroll round your garden every day or so, inspect your plants, and ensure they all receive a regular health check.

Lack of water is a common cause of problems with plants grown in containers. Bamboos are notoriously thirsty and need regular watering to keep them healthy.

WHAT'S WRONG?

Recognizing the early warning signs of plant disorders is essential
for good pest and disease control. Plants have only a limited number of
responses to stress, attack, or injury, and so similar symptoms can result from
a range of different problems. Inspect sick plants carefully, don't jump to
a hasty diagnosis, and avoid reaching for the pesticide spray too readily.

PLANT DISORDERS

Plants that look as if they are troubled by pests
or have a disease, but are not visibly under attack,
may be suffering from a "disorder". Disorders happen
when a plant lacks sufficient nutrients or water, or is
growing under adverse conditions, such as excessive
cold or warmth. They can also occur if plants are
subjected to more wind or sun than they can handle,
are growing in shade when they prefer sun, or have
a poor root environment.

Disorders are very common, and if the underlying
problems are not rectified quickly, they often lead to
an outbreak of disease or attack by pests, which take
hold of the weakened plant. Plant disorders are not
diseases, although the symptoms can look similar.
Adverse weather, such as drought or unusually low
temperatures, can affect plant health, although
mature, hardy types tend to recover quickly when
more favourable conditions return. Young plants,
or those that have recently been planted, are more
vulnerable; you will need to keep a close watch on
them, and protect them when necessary. Increase
their natural levels of resistance by making sure you
are growing them in a suitable site and soil.

Drought starves a plant's roots of
moisture and can cause its leaves to
develop brown edges – as in the case
of this primrose.

Overwatering can do as much damage
as underwatering. Make sure your soil
has good drainage to prevent the roots
from drowning.

Yellowing young leaves often indicate
an iron and manganese deficiency in
acid-loving plants, such as rhododendrons,
when they are grown in alkaline soil.

Sun scorch may cause leaves to turn brown. These Japanese maples prefer a shady site to being in full sun.

Frost damage may occur on the flowers of plants that bloom in early spring, even if the plant is hardy.

Lack of light causes plants and seedlings to grow tall and thin, and leaves to turn yellow. This is known as "etiolation".

LACK OF WATER OR NUTRIENTS

Lack of nutrients is a less common problem than you'd think; most tended garden soil has been fed over many years, topping up nutrient levels. Nitrogen, however, may be lacking, since it is easily washed out of the soil. A deficiency will lead to pale leaves and sickly growth, and can be remedied using a nitrogen-rich fertilizer, such as dried chicken manure pellets or sulphate of ammonia, and watering well. Organic matter, such as compost or manure, dug into the soil before planting or used as a mulch, will also boost nitrogen levels.

A lack of water causes poor growth and flowering, as well as powdery mildew attacks. Without an adequate water supply, plants cannot take up enough nutrients from the soil. During prolonged dry spells, plants may go brown and lose their foliage. Well-established woody plants have enough roots to ride out these dry spells and recover when rain returns. Lawns go brown, but the living buds at the base of the leaves remain alive, so they regrow when rehydrated. But plants with limited root systems, such as annuals and herbaceous perennials, are more vulnerable; cut back wilted stems, remove dead leaves, then water the plant well regularly. You should soon see new buds and stems emerge.

Newly planted plants, especially trees, shrubs, and climbers, are vulnerable to drought and a lack of nutrients. To prevent problems, water them regularly, and keep the area around the roots free of weeds.

LACK OF LIGHT

Growing plants in shade that prefer a sunny site will lead to pale or yellow leaves and stems, and poor growth. Plants and seedlings may grow tall and spindly, and lack vigour. Move the plants to a location where they will receive the right amount of light.

ROOT PROBLEMS

When soils are waterlogged, dry, or airless, roots are more likely to contract fungal diseases. The first signs of root problems may be in the foliage; when roots are too dry, the leaves furthest from the roots go brown, especially at the edges. Drought often causes the bark at the base of the stem to rot and die, too, although this is also a symptom of the root disease honey fungus (see *p.322*). If the base of the stem is sound, check for dead roots that appear red or brown and brittle. If the root system is dead or damaged, then the plant's chances of survival are slim, however if roots are healthy, it is possible that watering may save the plant.

GARDEN PESTS AND PARASITES

Pests are creatures that prevent gardeners from enjoying their garden, usually by damaging cultivated plants. Soiling by cats and foxes, and furniture and paving spoilt by sticky honeydew are other forms of damage. Pests may be snails, slugs, mammals, birds, mites, and, most commonly, insects.

PICKY EATERS

Most beetles and caterpillars are picky about what they will eat. Viburnum and lily beetles, for instance, stick to their respective hosts. Most greenfly and blackfly, officially called aphids, also have a limited range of host species. The trouble is that there are so many different kinds that aphids of one sort or another will attack most plants at some time. These insects obtain nutrients by sucking plant sap through their thin, hollow mouthparts. Other sap-suckers include thrips, scale insects, and capsid bugs. To aid feeding, some sap-sucking insects inject materials that regulate plant growth, and induce severe distortion and curled foliage that hides and shelters the insects. Mouthparts may also spread viruses.

GARDEN GLUTTONS

Slugs and snails eat most plants; seedlings, herbaceous plants, and climbers are worst affected. Holes appear at the edges of and in the middle of leaves, and you may notice telltale slime trails. Slugs also bore into bulbs and tubers. They feed in mild, humid weather, mainly at night. When slugs and snails damage soft growth, other problems often follow.

Lily beetles are bright red with black heads. Both adults and grubs eat leaves, flowers and seed pods, and may kill plants unless removed or treated.

Green aphids, or greenfly, excrete sticky honeydew on leaves and stems, providing an ideal growing medium for the black fungus known as sooty mould.

Earwigs cause little harm, despite their fierce appearance – they mainly take bites out of leaves and petals. They also benefit gardens by eating insect pests.

Vine weevils attack a wide range of plants. The adult insects, all female, cannot fly; they crawl around the garden from mid-spring to mid-autumn, usually after dark, laying eggs. They feed on leaves, mainly of herbaceous plants and shrubs (especially evergreens), notching the leaf edges only, as their mode of feeding is not adapted to making holes in leaves. The eggs, too small to be seen without a hand lens, hatch into grubs, which feed on roots. The grubs start to cause damage in late summer and early autumn, which is when you should apply controls.

CONTROLLING PESTS

Acting swiftly when problems arise eases their control, so watch for signs of trouble. Often, pruning out diseased shoots or removing cover for harmful pests will put a stop to damage. At other times, pesticides may be needed; you will only need a small amount if a problem is dealt with quickly. Ready-to-use packs are useful for small infestations – they save the chore of making up a small quantity of the solution.

Organic gardeners will only use pesticides that are derived from natural sources, such as pyrethrum. Although they may be less effective than synthetic products, they are good enough for most small insect pests. In fact, when used in the early stages of a problem, they may be all you need to stop the trouble before it gets out of hand. If you want to take an organic approach, it's also worth considering biological controls (see *p.327*). Pest controls work in a variety of ways:

• **Systemic controls** are absorbed into the plant's sap, killing fungi and sap-sucking insects as they feed. Thorough coverage with a spray is less important with these pesticides than with contact-action controls.

• **Contact insecticides** include the naturally derived materials, as well as some synthetic pesticides. The chemical has to touch the pest to do its job, so you must cover the upper and lower leaf surfaces. Act quickly: once foliage becomes distorted and curled it is difficult to get contact materials where they are needed.

• **Dusts** are contact materials and must be applied early. Puffer packs are not very satisfactory applicators and sprays are usually preferable to dusts.

• **Pellets and baits,** such as slug pellets, have a bad press and can be harmful if misused. If you follow the instructions, though, and store them safely and securely, they are a useful and safe treatment.

Vine weevils can be introduced into your garden in the roots or top growth of newly-bought plants. Always check plants and rootballs before planting.

Vine weevil larvae in roots often kill pot plants. Control in late summer with insecticides watered onto the pot, or use the nematode biological control.

Pollen beetle populations surge in midsummer. They do little damage to plants, but are unsightly on cut flowers; they can simply be shaken off.

IDENTIFYING PEST DAMAGE

Pests are often identifiable by the damage they cause. Many are elusive; some, like mites and eelworms, are microscopically small. Others fashion hiding places for themselves within leaves, while some, such as the gall formers, secrete chemicals that induce the plant to "grow" them a home. Many pests cover themselves with protective coatings, such as silk webbing, or the "cuckoo spit" that contains froghoppers, rendering pesticide sprays less effective.

1 Leaf eelworms kill leaf tissues and spread in water on leaves. Clearing away damaged foliage reduces their spread. Replace badly affected plants.

2 Galls may be unsightly but in most cases cause no serious damage to plants. Removing affected leaves on lightly infested plants may reduce subsequent attacks.

3 Caterpillars can strip leaves bare. Some are obvious, living in colonies, often beneath webbing. Others feed on plant roots, or work at night, hiding under foliage by day. Use contact insecticides when picking off doesn't work, and dig out soil-dwelling caterpillars when they reveal themselves.

4 Slug and snail damage is most common on vulnerable plants like this hosta; they will need protecting with slug control materials, or with biological controls.

5 Leaf rolling sawfly grubs live within rose leaves that roll up around them. Picking off and destroying affected leaves before midsummer may limit attacks the following year.

6 Snails' shells make them less dependent on moist cover than slugs; you may therefore find them on climbers, or on bulb leaves and flowers, such as this hyacinth.

7 Capsid bugs are seldom noticed until it is too late. Insecticides applied when the damage is in its early stages may help, as can good weed control.

8 Vine weevils feed on the leaves of a number of different plants, such as this rhododendron. They are particularly common where "wild" rhododendrons grow.

9 Sooty mould is often the result of an infestation of aphids. These pests excrete sticky honeydew which provides an ideal growing medium for the black sooty fungus.

10 Leaf miners are quite common, but generally not especially harmful pests. There are many different kinds; those found on primulas (**10a**) are different from those that attack holly (**10b**).

PLANT DISEASES

Most plant diseases are caused by fungi, others by bacteria or other micro-organisms, or viruses. They sometimes enter a plant through a wound; once inside, they grow and multiply and the plant sooner or later shows visible symptoms, and may die. Gardeners often spread plant diseases inadvertently when they acquire infected material, so always refuse offers of suspect plants.

Keep your garden equipment clean to reduce the risk of any problems spreading. Sterilizing and scrubbing pots, canes, posts, and wires will help prevent carry-over of spores and eggs to new areas of planting.

FUNGI

Fungi that infect plants may consist of microscopic threads; these can grow together to form structures that are visible to the naked eye. The first signs of infection that you are likely to notice are usually spots, rots, or stem dieback. Alternatively, it may be possible to spot signs of mildew, mould, or other evidence of fungal disease.

Typically, airborne fungi produce spores that spread by air currents, rain or watering splash, or, less importantly, on insects or seeds. Spores need moisture to germinate, which is why fungal diseases – mildews and moulds, for example – are so much worse in wet weather.

Many fungi also produce resting bodies that are able to survive dry or cold periods. These persist in the soil until better conditions or the presence of a susceptible plant is "detected". After this, they release spores that start the disease cycle over again. Fungi also persist in infected plants. Dieback diseases carry over from year to year in infected shoots and rose black spot will lurk in fallen leaves ready to reinfect roses in spring.

BACTERIA

Bacterial infections can cause a range of symptoms. Rots of soft tissue, such as bulbs and rhizomes, with a distinctive evil smell, are often caused by bacteria. Cankers, such as bacterial canker of cherries, are also widespread. Another common bacterial disease is fireblight, which affects plants in the apple and pear family, including pyracantha and cotoneaster. The symptoms include dieback of branches, and leaves that hang as if scorched by fire. Fireblight may be spread by insects that carry the bacteria from flower to flower, and like other bacterial diseases may be spread by air currents, or by rain or watering splash. Many diseases need help to enter plants, and gain access through insect pest damage, pruning wounds, or even the scars left from leaf fall in autumn.

VIRUSES

Viruses may spread when insects carry them between plants. Greenfly or aphids are common carriers, but thrips, whitefly and certain other insects play their part. Viruses are widespread in weeds and in most mature garden plants, where their effect is usually not noticeable – robust, mature plants are often able to stand up to their effects.

PREVENTING AND TREATING DISEASES

Contact fungicides form a protective layer that prevents spores from getting into the plant, but they will not cure infections. Cover the whole plant before the disease gets a grip on it. Prompt application is especially important for young plants, so inspect them regularly. If you leave it too late, they can die or are so slowed up in growth that they never make good plants. Viruses may be spread by insects, so controlling the insect will help in controlling the spread of the virus.

USING CHEMICALS

Reading and following the instructions on the label is vital for safe, effective, and legal pesticide use. Calculate how much you need. If in doubt, make up less; you can always mix more if it is needed. If only a few plants need treatment, use a ready-diluted spray formulation. Take care to avoid spillages and never allow pesticide to get into drains, ponds, streams, or ditches. Set nozzles to give a fairly fine spray that neither drifts nor drenches, and apply the spray evenly over the affected plants. Foliar sprays should not be applied so heavily that dribbles occur, but just enough to evenly wet the target.

(top) **Spray both the upper and lower surfaces** of foliage. Remember to keep pets and children away from the sprayed area until the material has dried onto the leaves.

(bottom) **Store pesticides** in a secure, locked cupboard in a cool place, and ensure they are clearly labelled; never decant pesticides into food or drink containers. Only use local authority facilities to dispose of old or surplus chemicals.

IDENTIFYING PLANT DISEASES

Canker, honey fungus, coral spot, and bacterial canker are all extremely serious; there are some sprays for canker but usually all that can be done is to cut away the affected parts, or remove the plant. Seedlings, too, will not recover from damping off disease and an affected batch must be disposed of. Virus-affected plants may also have to be removed, although surprisingly, they often produce healthy seed. Other diseases may be treatable, but in all of these cases, never add any affected plant material to the compost heap.

1 Canker in apples and pears makes shrunken, cracked rings of dead bark appear; it is caused by a fungus. Cut out all affected wood as soon as you spot it.

2 Honey fungus is a destructive, incurable fungus that spreads from plant to plant by dark strands, seen here beneath the bark. Affected plants must be removed.

3 Hollyhock rust and other types of rust disease disfigure rather than kill plants. Choose resistant varieties, and keep plants in good health to help resist attacks.

4 Bacterial canker, such as on this cherry tree, is a bacterial disease; copper may be used to control it. If you cannot cut out all infected wood safely, you will have to remove the tree.

5 Coral spot causes pink/red spots that dot the bark of dead wood, but it can invade healthy shoots of many trees and shrubs. It must be cut out.

6 Rose mosaic virus attacks a number of plants, causing different symptoms in each. Badly affected plants have to be replaced.

7 Rose rust spots (here, in spring) shed fungal spores in the summer, and this is followed by leaf loss. Rusts are a large group of diseases, characterized by orangey spots or "bumps".

8 Grey mould – also known as botrytis – often affects damaged plants. Careful handling and good slug control minimize the risk. Flowers, as well as leaves, can be affected (**8a**). Crowded plants grown indoors in a stuffy atmosphere are very prone to grey mould (**8b**).

9 Silver leaf disease shows as a silvery sheen on the leaves of fruit trees, such as plum and cherry trees. It attacks the tree through open wounds or cuts, particularly during winter when healing takes longer. To reduce the risk, prune either in early spring or just after the fruits are harvested. Silver leaf develops branch by branch – if the whole tree is affected this is false silver leaf, a physiological disorder, from which the tree will recover.

10 Rose black spot is a fungal disease that is prevalent in wet weather, spreading by water splash in wet windy conditions. Regular treatment with a fungicide is required, but removing infected leaves and pruning out diseased wood also helps. Catalogues list roses that are more resistant to this disease than others; while these are worth considering, no rose is completely immune.

11 Diseased fruits and berries can persist on trees or the ground over winter, ready to release spores in spring. To prevent this, put them in the dustbin, on a bonfire, or in a deep hole.

12 Damping off affects seedlings at soil level, making them collapse. Proper hygiene reduces the risk of this occurring.

13 Powdery mildews (**13a**) are specific to their host plants. Fungicides are effective at controlling mild attacks. Powdery mildew thrives on plants growing in the dry soil close to walls, such as clematis (**13b**). Mulching and watering reduces plants' vulnerability.

KEEPING PLANTS HEALTHY

Plants, like people, are much less susceptible to disease if they are
living in naturally healthy conditions than if they are under stress or
malnourished. Most problems occur when plants are neglected or denied what
they need. It's common sense, then, that the best gardening techniques are all
about giving plants what they require to grow and flourish successfully: a sheltered,
uncrowded site with the appropriate amount of light and warmth, adequate soil
preparation, sufficient water and nutrients, good drainage, and perhaps also
the presence of beneficial wildlife to help control insect and other pests.

Damp, shady woodland gardens look lush when planted with
big-leaved, deep green foliage plants that like the moisture.
Plants that hanker for sun and whose roots resent being wet
would soon fade away and succumb to pests and diseases.

BARRIERS AND TRAPS

Although barriers and traps require some effort to install, in many cases they provide full protection against pests.

• **Large mammals** can be fenced out of the garden. Fences for deer should be at least 1.8m (6ft) high all around the garden; rabbit barriers must be 1m (3ft) or more high and sunk a further 30cm (1ft) beneath the ground to prevent rabbits burrowing under them.

• **Birds** will not be able to feast on your fruit trees and bushes if you surround them with fruit cages. Although these barriers can be costly, they do provide peace of mind.

• **Small mammals** can be trapped, but this is often best left to the professionals.

• **Slugs and snails** are too well adapted to surmount obstacles and RHS research has found barriers ineffective in stopping them. However, bands of copper around the top of a pot or in beds and borders or rings of sharp materials that slugs and snails may find scratchy to cross can be tried around vulnerable plants. Grit, crushed eggshells, proprietary granule barriers, and cocoa shell mulch are also claimed to give protection.

• **Ants** and other pests can be prevented from climbing prized plants by collars of aluminium foil turned outwards; these need refreshing as the plants grow. Bands of a plant-friendly grease can be applied to the stems of trees in the autumn to prevent wingless pests, such as female winter moths, ascending. Fruit trees, including crab apples, are most likely to benefit from this. If the tree is staked, you must grease the stake too, or it will provide a bypass for pests up into the crown of the tree.

NATURAL ALTERNATIVES TO TRAPS

If you don't like the thought of trapping the creatures in your garden, "companion planting", which helps to deter them, is always an alternative option that you can consider. For example, not only are marigolds (*Tagetes*) said to keep flying insect pests off plants such as roses, but their root secretions are also thought to help prevent soil-borne diseases and pests, such as nematode worms.

(top) **Slug and snail traps** range from the homespun – a hollowed out potato, or the shell of half a grapefruit upended on the soil – to this hi-tech version of the beer trap.

(bottom) **Companion planting** is a very old but largely unproven method of natural pest control. Traditional companion plants such as these marigolds may deter flying pests, including whitefly, and their root secretions are thought to inhibit soil pests and weeds.

GOOD GARDENING FOR HEALTH

Ward off pests and diseases in your garden by following a few simple rules:

• **Protect plants** against the cold, wind, and drought; they are especially vulnerable when newly sown or planted. Spread out roots and carefully secure them in well-prepared soil, avoiding excessively deep planting. Faults here may be fatal, but not before years of disappointing growth.

• **Rotate bedding plants** to avoid a build-up of pests and diseases; don't plant them in the same place in consecutive years.

• **Weed regularly** to reduce pest and disease problems. Clean up fallen leaves, fruits, and dead wood. Do not compost or recycle infested material; remove it by burning it or putting it in the household waste.

• **Avoid damage** to plants – often caused by mowers, strimmers, and by pruning – by using tree guards and edging strips on lawns. Damage opens up the way for diseases and rots.

• **Avoid dense planting** to increase airflow between plants; thin any crowded growth by pruning. The lower humidity levels should help reduce moulds and rotting. Indoors, too little humidity is more likely, encouraging red spider mite and whitefly attacks. It can be countered by misting and dampening plants frequently, and by standing plants in trays of moist gravel or inside larger pots of damp moss.

• **Avoid overwatering,** as permanently wet roots can lead to root rots and dieback, whereupon even mildly harmful fungi, such as coral spot, can cause great damage.

• **Avoid underwatering,** as dry roots and irregular watering weaken plants, encouraging powdery mildew.

A garden pond will encourage a range of beneficial wildlife. Make a "deep end" 60cm (24in) in depth for aquatic creatures to escape freezes, and a "beach", so small animals and birds may approach the water safely.

NATURAL ALLIES

Most insects are not pests – some, often seen in the vicinity of damaged plants, actually only take advantage of damage that has been made by the weather or by another pest. Some insects only cause problems under certain circumstances:

• **Ants** are a nuisance in lawns and may loosen the plants' grasp on the soil; counter this with watering, raking, and soil firming. They also cause problems when they "farm" aphids for their honeydew, protecting them aggressively from predators so that large and harmful colonies build up on plants.

• **Woodlice, centipedes, and millipedes** are excellent "recyclers" of plant debris, but might occasionally nibble at holes made by slugs and other pests, or damage delicate seedlings. Encourage them to go elsewhere by clearing away debris.

• **Bees** are vital for pollination, and therefore are among the creatures that should always be spared – even the leaf-cutting bees that take an occasional bite from plants.

• **Wasps** can be a nuisance throughout late summer, but before that they will obligingly prey on other insects. You should only destroy their nests if you really have to – for example, in a family garden.

BIOLOGICAL CONTROLS

Instead of using chemicals, release nematodes, insects, or mites that prey on certain pests. These biological controls are most effective in greenhouses and conservatories; when used outside they are likely to dissipate. When using natural predators to control pests, don't also use synthetic insecticides, and ideally only use those based on fatty acids or plant oils that leave no residues.

Biological controls for vine weevils and slugs are watered onto the soil or compost. For slugs, apply from spring to early autumn. The control for vine weevil is best used in containers in late summer before the grubs do much damage. Both work best on open-structured soil. Controls for red spider mite and whitefly are effective in the greenhouse as long as they are used before the pest becomes too numerous.

It's usually best to buy biological controls by mail order. A few garden centres stock them, but as they are living creatures, they are not easy to store.

ENCOURAGING PEST PREDATORS

Attract beneficial creatures into your garden by providing them with food and shelter. Features to tempt wildlife are generally inexpensive, many are attractive, and those that are less so are fairly unobtrusive. Gardeners can also create a wild corner of the garden where grasses and plants are left to grow unrestrained and allowed to stand over winter. Include a range of native species, and only use pesticides when and where they are really necessary.

Nesting boxes are far more likely to attract occupants if you also provide a feeding station nearby for birds to supplement scarce natural resources in winter.

Attract butterflies and bees with flowers rich in nectar and pollen, such as sedum and buddleja, and with open blooms that insects can get into easily.

Lacewing houses can be erected for these helpful insects whose larvae prey on a range of harmful garden pests, especially aphids.

Old logs piled in an undisturbed, cool area of your garden will provide a hiding place for helpful creatures, such as toads, frogs, newts, and beetles.

GARDEN YEAR
PLANNER

———

To make life easy, this garden planner shows you at-a-glance what jobs need to be done and when. Simply look at the season in question and follow the advice given – you will find more information on each subject in the relevant chapters in the book. For example, early spring is a good time to start preparing your soil for planting and there are further details on soils in the chapter starting on p.8. Winter is perhaps the best time to observe a garden critically – if the structure works in winter, it should work throughout the rest of the year with the distractions of flowers and foliage.

Here, astrantias and dark purple roses mingle with pale pink persicaria flower spikes and heuchera leaves for a beautiful summer display. Careful planning ahead will help you group together plants that flower at the same time of year.

EARLY SPRING
FEBRUARY–MARCH

This is the beginning of the gardening year, and heralds a busy period for gardeners. However, don't be over-eager: keep off lawns, beds, and borders if they are frozen or wet, as walking on them risks compacting the soil. And don't attempt to dig beds until the weather improves.

Prune buddleja

PREPARATIONS
• Start weeding: pull small seedlings by hand but dig out the roots of perennials such as dandelions, or use a systemic weedkiller.
• Prepare the soil for planting by forking it over and digging in organic matter.

SOWING SEEDS
• Sow seeds of half-hardy and tender bedding plants indoors in a frost-free place.

PLANTING
• Plant summer-flowering bulbs, hardy perennials, and ornamental grasses.
• Plant bare-rooted shrubs and garden trees.

PRUNING
• Prune bush roses.
• Prune deciduous shrubs and climbers that flower later in the summer, such as buddleja, jasmine, lavateras, and certain varieties of clematis.
• Prune hardy evergreen shrubs and trees.

OTHER TASKS
• Lift and divide over-sized clumps of perennials; discard dead or unproductive sections in the middle of the clump.
• Prepare soil for planting in fair weather.
• Rake the lawn, removing leaves and other debris, and brush off worm casts if you need to mow.
• Repot or top-dress shrubs, climbers, and alpines growing in pots and containers.

MID- TO LATE SPRING
APRIL-MAY

The garden is changing daily now, many spring bulbs are in flower, trees are in blossom, and spring-green foliage is everywhere. For many gardeners, this is the best time of the year. Yet the weather can still be unpredictable. April is prone to sudden frosts and even in warm areas they are not unknown in May. Don't sow seeds outdoors too early.

PREPARATIONS
• Apply a general-purpose fertilizer around established shrubs and roses.
• Mulch the soil after it has rained or been watered.
• Stake herbaceous perennials as growth emerges.

SOWING SEEDS
• Sow seeds of half-hardy and tender annuals inside, or buy plug plants and grow on in a frost-free place.
• Harden off bedding plants sown earlier in spring.
• Sow hardy annuals in situ.
• In late spring, thin hardy annuals sown outside.

PLANTING OUT
• Plant evergreens, and move any that are in the wrong place.
• Protect vulnerable plants by covering them or bringing them indoors at night.

PRUNING
• Prune early-flowering shrubs and climbers after they have bloomed.
• Cut back ornamental grasses as they start to grow.

OTHER TASKS
• Make new shrubs and climbers by layering.
• Start taking softwood cuttings from new shoots.
• Turf or sow new lawns, and repair worn-out lawns.
• Apply a spring fertilizer, and weedkiller, to lawns.
• Cut the grass with the blades quite high.
• Keep weeding.

Stake tall perennials as growth emerges.

EARLY SUMMER JUNE

Rising temperatures should now be coaxing summer flowers into bloom – roses and peonies come into their own, and perennial flower borders will soon be looking their best. You should have completed most of your seed sowing, pricking out, and potting on. And with the warm weather and the end of night-time frosts, tender plants can be moved or planted outdoors.

Plant up hanging baskets

PREPARATIONS
• Feed flowering shrubs and roses with a rose fertilizer to promote flowering.
• Weed carefully to avoid damaging nearby plants.
• Buy bedding plants and plant out after the frosts.

PLANTING OUT
• Plant up hanging baskets and summer pots outside when all risk of frost is over.
• Plant out half-hardy and tender bedding plants when the frosts are over.

PRUNING
• Prune late spring-flowering shrubs and climbers after they have bloomed.
• Trim vigorously growing hedges, e.g. privet.

OTHER TASKS
• Remove suckers from roses.
• Tie in climbers regularly to their supports.
• Divide congested clumps of primulas and irises after they have flowered.
• Continue to take softwood cuttings.
• Mow lawns regularly, and lower the blades.
• Water new plants in dry spells.

MID- TO LATE SUMMER
JULY-AUGUST

The work in the garden gradually eases off as summer progresses, leaving you free to sit back and enjoy the fruits of your labour. The likelihood is that you will spend most of your time watering during dry spells, deadheading plants that have already flowered, keeping beds and borders weeded, and pruning back the uncontrolled growth of your more over-enthusiastic and vigorous plants. If you have a corner of the garden devoted to vegetables, herbs, and fruit, now is the time to harvest and enjoy them.

PREPARATIONS
• Prepare the soil for a new lawn or border.
• Plant early-spring-flowering bulbs – such as snowdrops, crocuses, and daffodils – in late summer.
• Take semi-ripe cuttings of shrubs.

PRUNING
• Deadhead flowering plants to promote new blooms.
• Cut back herbaceous plants after flowering for a second burst of foliage and flowers.
• Prune rambling roses as soon as they have flowered.
• Prune established climbers and deciduous shrubs that flower on the previous year's growth.
• Trim evergreen and conifer hedges in late summer.

OTHER TASKS
• In dry spells continue to water new plants and those still establishing themselves, as well as new lawns. Water in the early morning and evening rather than in the heat of the day.
• Water and feed hanging baskets and other containers regularly.
• Feed and mulch roses.
• Continue to weed regularly.
• Mow lawns regularly – perhaps as often as once a week or more – but set the blades higher during hot, dry spells.

(above) **Mow** lawn

(below) **Water** containers regularly

AUTUMN
SEPTEMBER-NOVEMBER

As rain returns and the weather cools, get ready for a season of activity – planting hardy perennials, shrubs, and bulbs for the following spring, lifting and dividing established plants, clearing away the dead and dying plant debris of the summer, and protecting tender plants from the coming harsh conditions.

Plant spring bulbs

PREPARATIONS
• Remove dead and dying stems and foliage from perennials, but leave a few to protect the roots over winter.

PLANTING OUT
• Plant hardy perennials and hardy evergreen shrubs early in the season.
• Plant deciduous shrubs and trees in late autumn when the leaves have fallen.
• Remove and compost summer annuals.
• Plant spring-flowering biennials, such as wallflowers.
• Plant spring-flowering bulbs but leave tulips until November.

OTHER TASKS
• In late autumn transplant deciduous shrubs.
• Lift and divide hardy perennials.
• Lift and store, or insulate, tender perennials and shrubs before the frosts.
• Lift and store dahlias and cannas after the first frosts.
• Take hardwood cuttings of deciduous shrubs.
• Check tree ties and loosen if tight.
• Cover ponds with nets to keep out falling leaves which contaminate the water as they decompose.
• Apply an autumn lawn fertilizer.
• Lay turf or sow new lawns.
• Mow lawns less frequently as growth slows.
• Rake up leaves and debris from lawns.
• Aerate lawns using a fork and top-dress to stimulate root growth and prevent moss.
• Repair lawns, if necessary.

WINTER
DECEMBER-JANUARY

Cold winter months are a time of rest for most plants, as their growth cycles slow right down and they enter a period of dormancy before coming to life again the following spring. But there is still plenty to do in the garden – and venturing out in the fresh air can really raise the spirits.

PREPARATIONS
• Dig organic matter, such as well-rotted manure or garden compost, into clay soils.
• Protect slightly tender perennials with straw or compost.
• Brush heavy snow off trees, hedges, and shrubs to prevent the weight snapping branches.

PLANTING OUT
• Plant deciduous trees, shrubs, and hedges on dry, frost-free days, unless the soil is very wet.

PRUNING
• Prune trees (except plums and cherries) and wisteria, and renovate overgrown or old deciduous shrubs.
• In late winter, coppice shrubs for special effects like colourful stems or bold foliage.
• In windy gardens, prune back the top growth of tall roses and other shrubs to prevent windrock.

OTHER TASKS
• Check stored dahlias, cannas, and gladioli, and remove any that show signs of rotting.
• Wrap up pots with bubble wrap and cover vulnerable plants with fleece if frost is forecast.
• Continue to take hardwood cuttings of deciduous shrubs.
• Take root cuttings of perennials.
• Continue to rake up leaves from lawns.
• Keep off lawns and soil when sodden or frozen.
• Check and repair garden tools.
• Make plans for next year's planting.

Renovate deciduous shrubs

GLOSSARY

Acid Refers to soil with a pH of less than 7 and determines the range of plants that can be grown.

Aerate Digging or forking the soil or spiking a lawn to allow more air in.

Alkaline Refers to soil with a pH of more than 7 and determines the range of plants that can be grown.

All-purpose fertilizer See Fertilizer.

Alpine Like "rock plants", often applied to a wide range of small plants, but more correctly, from mountainous areas.

Annual A plant that germinates, flowers, sets seed, and dies in one growing season. Hardy annuals can be sown the autumn before; half-hardy ones are sown in spring.

Aphids Greenfly or blackfly.

Architectural Buzzword used to describe plants with a strong outline or bold leaf shape.

Aspect Which way a garden, a wall, or a plant faces.

Bare-rooted Nursery-grown trees and shrubs, dug up and sold without a pot or any soil around their roots.

Bedding Temporary planting for seasonal displays.

Biennial A plant that grows one year, flowers the next, and then dies.

Biological control Deliberate release of a tiny organism that kills certain garden pests.

Bleed To lose sap freely through a cut or wound.

Blind A shoot or a bulb that fails to flower.

Blood, fish, and bone See Fertilizer.

Bract A modified leaf at the base of a flower or cluster of flowers. Bracts may be colourful like petals.

Brassica A member of the cabbage family.

Broadcasting Scattering seed or fertilizer evenly over the soil, rather than distributing it in drills or furrows.

Bulblet A baby bulb produced from the base of a mature bulb, and easily separated from it.

Cell tray See Module.

Cloche A low, clear plastic or glass dome, tunnel or box used to protect young garden plants, or to warm the soil.

Cold frame A glazed, box-like, unheated structure with a lid, used to protect plants from the elements.

Coir Recycled outer fibres of coconuts, used as an ingredient in potting composts.

Compost 1. Potting medium made up of any combination of soil, sand, peat or peat substitute, leafmould, or other ingredients. It may be "multi-purpose", for general use, or tailored for certain plants, for example, cuttings or alpines. 2. A home-made organic material formed from plant waste and other organic matter, used to improve soil or as a mulch.

Compacted soil With all the pockets of air that plants need squashed out of it, e.g. after heavy use of a lawn.

Composted bark By-product of the timber industry that makes a good soil conditioner. Not to be confused with chipped bark, a mulching material. Bark chips dug into the soil rob it of nitrogen and may take years to decay.

Containerized Plants raised in the open ground, potted up for sale.

Coppicing To prune back a tree or shrub close to the ground to stimulate growth of young stems.

Corm An underground bulb-like body. When mature, baby "cormels" or "cormlets" form around it (see also Bulblet).

Crocks Shards of broken terracotta pot, or similar, placed at the bottom of containers to improve drainage.

Cross-pollination When pollen from one plant fertilizes the flower of another.

Acid-loving rhododendron

Annual

Colourful bracts above flowers

Brassica

Hard landscaping

Hardy plant (rosemary)

Green manure

Herbaceous plant

Crown The base of a herbaceous plant where the stems join the root system, from where new shoots grow.

Cultivar Cultivated varieties of plants selected for desirable characteristics, such as flower or leaf colour. Their names are written within inverted commas.

Cuttings Parts of a plant – e.g. sections of stem or root – th.at can be grown on to form a new plant.

Damping off Very common disease that causes small seedlings to keel over at soil level and die.

Deciduous Of trees and shrubs that drop their leaves every year in the autumn.

Division A method of increasing perennial plants by dividing them into sections, each with roots and one or more shoots.

Dormancy When growth temporarily ceases or other plant functions slow down due to seasonal effect of low temperatures in winter, or other adverse conditions, like excessive heat or drought. Seed dormancy prevents germination until conditions are favourable.

Double-digging Deep digging, turning the soil over to a level equal to two spade-blades ("spits") in depth.

Drill A narrow, straight, shallow furrow or groove into which seeds are sown or seedlings planted.

Ericaceous compost Acidic compost, suitable for growing lime-hating plants in pots, like heathers (*Erica*).

F1 hybrid Plant bred for uniformity and desirable flowers or other characteristics. Seed from F1 hybrids will not come true.

Fertilization Successful pollination resulting in the formation of seed.

Fertilizer A source of plant nutrients. Organic fertilizers include blood, fish, and bone, pelleted chicken manure, and seaweed-based products; inorganic fertilizers are chemical formulations. "General", "all-purpose", or "balanced" fertilizers serve most purposes. May be in granular, powder, pelleted or liquid formulations. Controlled-release fertilizer, in pellets or granules, releases its nutrients into the soil over several months.

Fleece A light, woven, translucent or transparent fabric that can be used to protect plants from cold or from pests.

Frost pocket A low area of land where cold air collects.

Garden compost *See* Compost (2).

Germination When a seed starts to grow into a plant.

Glaucous A blue-green, blue-grey, grey, or white bloom on leaves or other plant parts.

Graft union The point where a bud or stem is joined to a rootstock of a different plant when being grafted.

Green manure Fast-maturing crop, for example, clover, grown and then dug into the soil before it flowers to raise the soil's fertility.

Growing season Usually from spring to autumn when temperatures are above 6°C (43°F).

Half-hardy A plant that will not survive temperatures much below freezing.

Hardening off Gradually acclimatizing plants that have been raised indoors to cooler conditions outside.

Hardiness How resistant to cold a plant is.

Hard landscaping Anything in the garden that is not living plants, hedges, or lawn, e.g. paths, walls, gravel.

Hardy Plants that can withstand outdoor conditions year round and temperatures below freezing without protection.

Heavy soil With a high clay content, hard to work.

Heeling in Temporary planting to protect a plant until it can be placed in its permanent position.

Herbaceous A non-woody plant with upper parts that normally die down to a rootstock in the winter, usually applied to perennials, or borders planted with perennials.

Herbicide A chemical that kills plants.

Horticultural grit A fine, washed grit added to potting compost or heavy clay soil to improve its drainage. Also used as a mulch.

Horticultural sand Fine washed sands are used as a top-dressing for lawns and for propagation benches. Coarse washed sand can be added to soil and potting media to improve drainage. Builder's sands may contain materials harmful to plants and soils.

Humus The residue of organic matter in the soil that helps to improve it.

John Innes composts Soil-based proprietary composts made to standard formulas. No. 1 is for seeds and young plants; No. 2 is general-purpose; No. 3 is for shrubs and other long-term plants.

Legume (mange tout)

Marginal plant (iris)

Pergola

Pricking out

Landscape fabric *See* Weed-suppressing membrane

Lateral A side shoot or root.

Layering Method of propagation involving encouraging a stem to grow roots where it touches the ground, while still attached to the parent plant. Useful for climbers.

Leader, leading shoot The main or central shoot of a branch or a young tree.

Leafmould Rotted-down, fallen (i.e. dry) leaves. It is a good soil conditioner.

Legume Member of the pea or bean family, e.g. sweetpea.

Lime Horticultural product added to soil to raise the pH.

Limy soil Alkaline – e.g. chalky – with a pH above 7.

Light soil With a high sand content, fast-draining.

Loam Soil comprising ideal amounts of sand, clay, and silt, easy to work, water-retentive and free-draining.

Marginal A plant that can grow well in shallow water or in moist soil at the edge of a pond or stream.

Microclimate Spot in the garden with conditions that differ slightly from the norm, e.g. a sheltered, sunny wall.

Mixed border With both shrubby and herbaceous plants.

Modules or cell trays Used for sowing seeds in, consisting of separate compartments, or cells.

Mulch A material spread over the soil surface to suppress weeds and conserve moisture.

Mushroom compost The spent growing medium used by mushroom farmers; a useful soil conditioner and mulch but contains lime.

Native plants Originating in the country or area you live in.

Naturalistic Informal, imitating nature.

Naturalized Introducing plants so that they look as if they are growing naturally – for example, snowdrops in grass.

Neutral soil With a pH of 7, neither acid nor alkaline.

NPK Chemical abbreviations of the major plant nutrients, nitrogen (N), phosphorus (P), and potassium (K).

Nursery bed A small area of the garden set aside for sowing seeds, propagation, or growing on young plants.

Nutrients Minerals needed for healthy plant growth.

Offset A baby plant produced alongside a larger one, easily detached from the parent.

Organic matter Anything derived from plants or animals that is dug into or spread over the soil to improve it. May be the result of recycling – for example, garden compost – or leafmould, manures, or chipped bark.

Overwintering Moving non-hardy plants into a protected, frost-free place, such as a greenhouse or conservatory, to pass the winter months until late spring or early summer when frost is past and temperatures have risen.

Pea sticks Twiggy branches ideal as supports for tendril climbers, such as peas.

Peat substitute Container medium, such as composted bark, that provides similar conditions to peat. Using such products protects natural peat bogs, an important habitat.

Perennial A plant that lives for at least three growing seasons; usually applied to herbaceous plants.

Pergola A linked series of arches over a path or patio, either open at the top or with a roof.

Pesticide A substance used to kill pests.

pH A measure of acidity and alkalinity on a scale of 1 to 14: 7 is neutral, below 7 is acid, and above 7 is alkaline. You can buy pH testing kits from garden centres.

Photosynthesis The complex process by which green plants manufacture their own food, in the form of carbohydrates.

Pinching out Nipping out the tips of shoots with finger and thumb to make the plant bush out lower down.

Pollarding The regular or annual pruning back of stems to a short trunk to form a small, lollipop-shaped tree.

Pollination Transfer of pollen from one flower to another.

Pollinator The pollinating agent, e.g. bees; the wind.

Pot-bound A plant with crowded roots that has grown too big for its container.

Potting on Moving a plant from one pot to a larger one.

Pricking out The transfer of recently germinated seedlings that are growing closely together into pots where they have more room to grow.

Propagation Making more plants by raising seed, rooting cuttings, grafting or other means.

Propagator A device with a glazed cover and the means to regulate humidity and temperature for seed germination and the growth of young seedlings and cuttings.

Rain shadow The dry ground next to a wall or fence that is sheltered from prevailing winds and rainfall.

Reverted When the leaves of a variegated plant go back to being all green.

Rhizomes Branching stems with buds and roots growing close to soil surface. Sometimes fleshy, e. g. bearded irises.

Ripe/ripening Used to describe a shoot of a shrub or tree when it has become mature and firm over the course of a summer.

Rootball The roots and soil or compost that surrounds them in a container, or in the soil.

Rootstock 1. The roots of a plant used in grafting. 2. The root system of a herbaceous plant that survives over winter while the top growth dies down.

Runners Stems that lay on the soil surface and root at the tips – for example, strawberries.

Scarification The removal of debris from a lawn using a rake or other tool.

Seed leaves The first leaves a seedling develops.

Seep hose A hose pierced with holes to allow water to dribble out, very slowly, along its length.

Self-seeding The unaided natural scattering of seeds that produces seedlings around the parent plant.

Sheet mulch *See also* Geotextile. Also, a plastic sheet spread over the soil to warm it up before sowing seed. Black plastic sheeting can be laid over soil for up to a year to clear the ground of perennial weeds.

Shelter belt Trees or shrubs planted to protect the garden from wind, but taller and wider than a hedge.

Soil mark The dark mark on the stem(s) of an uprooted tree or shrub that shows where it originally met the soil.

Sour Foul-smelling soil or compost that lacks oxygen and contains stagnant water, resulting from waterlogging.

Specimen Plant with a striking shape that looks good planted on its own, e. g. a tree in a lawn.

Spit The depth of a spade blade, usually 25–30cm (10–12in).

Spore The minute reproductive structure of flowerless plants, such as ferns, fungi, and mosses.

Stress Plants under stress cannot function properly, start to wilt, and become more vulnerable to pests and diseases. The cause is usually lack of water due to dry (or frozen) soil, but may also be waterlogging or physical damage. If conditions do not improve, plants may die.

Subsoil The layers of soil beneath the topsoil: often paler and less fertile than the topsoil above it.

Succulent A plant with thick, fleshy leaves and/or stems that is adapted to store water, e.g. sedum.

Sucker 1. A shoot that grows from a plant's roots or an underground stem. 2. On grafted plants, a sucker is any shoot that grows from below the graft union, and should be removed.

Tamp To firm or pack down soil to bring it into contact with plant roots or newly-sown seeds.

Tap root One main downward-growing root, e.g. a carrot.

Tender A plant that will not survive temperatures below 5°C (41°F).

Tendril Modified leaf, branch, or stem, capable of attaching itself to a support.

Thatch A layer of dead material that accumulates on the surface of lawns.

Tilth A fine, crumbly layer of soil ideal for sowing seeds.

Top-dressing 1. A mix of compost, topsoil, and/or fertilizers applied around a plant or on a lawn to replenish nutrients. 2. A decorative layer, e.g. gravel laid over the soil.

Topsoil The upper layer of soil, darker and more fertile than the subsoil.

Transpiration The loss of water by evaporation from the leaves and stems of plants.

True Resembling the parent plant. Seed that will not "come true" will result in plants that may look different from their parent.

Variegated Usually leaves that are splashed or margined with a different colour.

Variety Everyday word used to describe a cultivar.

Vermiculite A lightweight mineral that retains water well and improves the texture of potting compost.

Wall shrub Any shrub trained and tied in to a wall.

Weed-suppressing membrane A man-made water-permeable material that suppresses weeds.

Whip A young tree, yet to grow any side branches.

Pot-bound

Sucker

Succulent

Variegated ivy

UNDERSTANDING PLANT NAMES

The plant world is infinitely varied, from tiny mosses to giant trees, and flowering plants to plumed grasses. To bring order to this diversity, plants are classified into groups on the basis of shared characteristics, such as the structure of the flowers and fruit, or the arrangement of the leaves. Plant names are an important part of the classification; you may think them unwieldy, but botanical names are used world-wide as a means of precise identification, which is not possible with common names. Common names are often very local: one plant may have several names, or one common name may refer to a few different plants; for example, the name "marigold" could be describing a *Calendula, Tagetes,* or *Caltha.*

Families (clockwise from top left) Rose (Rosaceae), Sunflower (Asteraceae), Lily (Liliaceae)

Genera (clockwise from top left) *Tagetes, Achillea, Aster*

Species *Helianthus annuus*   **Cultivar** *Helianthus annuus* 'Teddy Bear'   **Hybrid** *Helianthus* x *multiflorus*

Family

Plants with similar characteristics are grouped into families, e.g. daisy-, sunflower- or aster-like plants are in the daisy family (Asteraceae). Members of the same family are then broken down into smaller groups to reflect differences.

Genus (plural, genera)

The genus is like a surname; eg the genera in the daisy family (Asteraceae) include *Aster*, *Achillea*, *Tagetes*, and *Helianthus* (see *opposite*). But within each genus are plants with different characteristics, so further division is needed.

Species

This is like a forename, but it follows the genus; here, *Helianthus annuus* (see *above*) is the annual sunflower. There can also be some variation within a species that may have occurred in the wild, or as a result of plant breeding.

Cultivars

Cultivar (short for cultivated variety) names are given to plants that have been bred or selected for decorative or useful characteristics. *Helianthus annuus* 'Teddy Bear', for instance (see *top centre*), has been bred from the annual sunflower for its large, double flowers.

Hybrids

You may come across names with an "x" in the middle. This means the plant is a hybrid, the result of two different species cross-pollinating. In the case of *Helianthus* x *multiflorus* (see *top right*), *H. annuus* and *H. decapetalus* have cross-pollinated. This occurs both in the wild, and in cultivation.

BOTANICAL NAMES EXPLAINED

As well as understanding how a plant is grouped and identified, the name, or part of the name, often offers clues about it, too. Botanical names can be derived from Latin, Greek, or geographical terms, and may describe characteristics that make a plant distinctive. Here is a selection of common names.

• **Plant parts:** *baccatus* berry-like; *florus* refers to flowers; *folius* or *phyllus* refers to leaves.

• **Shape and habit:** *arboreus* tree-like; *elatus* tall; *fruticosus* shrubby; *pendulus* weeping; *scandens* climbing.

• **Colour:** *albus* white; *aureus* golden; *caeruleus* blue; *niger* black; *purpureus* purple.

• **Leaf shape:** *angustifolius* narrow leaves; *latifolius* broad leaves; *macrophyllus* large leaves.

• **General:** *armatus* thorny; *campanulatus* bell-shaped; *cordatus* heart-shaped; *fulgens* shiny; *mollis* soft; *officinalis* medicinal; *variegatus* variegated; *zebrinus* striped.

INDEX

Page numbers followed by a 'g' indicate an entry in the Glossary (*see pp.336–39*).

A

Acer (maple) 101
  A. campestre (field maple) 214
  A. griseum (paper-bark maple) 193
  A. japonica 'Aurea' 195
  A. palmatum (Japanese maple) 195
  A. p. 'Atropurpureum' 195
  A. p. 'Nicholsonii' 206
  A. p. 'Sango-kaku' (coral-bark Japanese maple) 20, 209
Achillea (yarrow) 22, 28, 41, 75, 283, 340
  A. filipendulina 'Gold Plate' 187, 191
  A. millefolium 179
acid soil 19, 20, 336g
Aconitum (aconite/monkhood) 35
  A. 'Bressingham Spire' 52
Actinidia 126
  A. kolomikta 128
Adiantum raddianum 'Gracillimum' (maidenhair fern) 165
aerating 172–23, 336g
Agave americana 55
Agrostemma githago (corncockle) 61
Ajuga reptans (bugle) 183
  A. r. 'Atropurpurea' 38
  A. r. 'Burgundy Glow' 52
  A. r. 'Multicolor' 65
Akebia quinate (chocolate vine) 129
Alchemilla mollis (lady's mantle) 25, 187, 191
alkaline soil 19, 21, 245, 336g, 338g
all-purpose fertilizer *see* fertilizers
Allamanda cathartica (golden trumpet) 131
Allium (onion) 34, 41, 90, 251

  A. cernuum (nodding onion) 96
  A. cristophii 22, 90
  A. karataviense 191
  A. moly 90
  A. 'Purple Sensation' 90
  A. schoenoprasum (chives) 262
  A. s. 'Forescate' 56
allotments 243
alpine plants 62–65, 336g
Alstroemeria ligtu hybrids (Peruvian lily) 17, 79
Amelanchier lamarckii (juneberry/snowy mespilus) 206
Anemone 92
  A. blanda var. *rosea* 'Radar' 76
  A. blanda 'Violet Star' 65
  A. pavonina 95
Angelica archangelica (Angelica) 262
annual plantings 29, 31, 32, 34, 69, 80–85, 230, 336g
  easy annuals from seed 84–85
Anthemis punctata subsp. *cupaniana* 45
ants 325, 327
aphids 316, 319, 336g
apples 264, 268
aquatic plants 304–05
Aquilegia formosa (columbine) 78
Aralia elata 'Variegata' (Japanese angelica tree) 206
Arbutus unedo (strawberry tree) 206
arches 123, 144
architectural plants 336g
Argyrocytisus battandieri (pineapple broom/pink Moroccan broom) 61, 116
Armeria maritima (sea thrift) 64, 283
Artemisia (mugwort/ wormwood) 187, 283
  A. alba 'Canescens' 43, 190

  A. 'Powys Castle' 56
  A. stelleriana 'Boughton Silver' 38
Artemisia dracunculus (tarragon) 263
Arum italicum subsp. 'Marmoratum' (lords and ladies) 76
arum lily 51
Aruncus (goatsbeard)
  A. aethusifolius 53
  A. dioicus 'Kneiffii' 78
Arundinaria gigantea subsp. *tecta* 158
Asparagus densiflorus 'Myersii' (foxtail fern) 165
aspect of garden 14, 17, 336g
Asplenium
  A. ceterach (rusty-back fern) 63
  A. scolopendrium Crispum Group (hart's tongue fern) 164
assessing your garden 8–25
Aster (Michaelmas daisy) 340
  A. frikartii 'Wunder von Stäfa' 45
  A. novi-belgii 'Lady in Blue' 79
Astilbe 75
  A. × arendsii 'Fanal' 52
Astrantia 329
aubergines 258
Aubreta deltoidea 64, 76
auto-irrigation systems 284
autumn 15, 35
  autumn colour 38, 113, 209
  bulb-like plants for 96–97
  garden tasks 334
  lawn care 172–73
  perennials for 77
  shade plants for 47
  shrubs for 101, 113
azara 17

B

bamboo 24, 149, 156–61, 313
  controlling spread of 158, 159
  cutting back 159

ornamental 160–61
  where to plant 156, 157
  see also individual species
bare-root plants and trees 265, 266, 336g
bark 286–87, 336g
beans
  French beans 256
  runner beans 256, 257
Beaumontia grandiflora (Herald's trumpet) 128
bedding plants 34, 67, 336g
  see also annual plantings
bees 296, 299, 327
beetroot 248, 250
Begonia
  tuberous 92, 93
  Begonia 'Herzog von Sagan' 97
Bellis perennis 179
Berberis (barberry) 29, 101
  pruning 219
  Berberis darwinii 'Barberry' 220
  B. thunbergii f. *atropurpurea* 'Helmond Pillar' 44
Bergenia (elephant's ears) 47
  B. 'Morgenröte' 78
  B. 'Sunningdale' 57, 76
berries 272
Betula utilis (birch) subsp. *jacquemontii* (Himalayan birch) 207
  B. u. subsp. *j.* 'Silver Shadow' 38
biennials 32, 69, 80–97, 336g
biological controls 296, 327, 336g
black spot 323
blackberries 272, 274
blackcurrants 264, 265, 275
Blechnum (hard fern) 163
  B. gibbum 165
bleed 336g
blind 336g
bog garden 51
Borago officinalis (borage) 262
borders 22, 26–67
  planting bulbs in 91
  size, shape, and layout 30–31

Brachyglottis 283

Brachyscome iberidifolia 'White Splendour' (swan river daisy) 230

bract 336g

brassicas 245, 255, 336g

Briza maxima (greater quaking grass) 155

broad beans 246

broadcasting 336g

broccoli 255

Brunnera macrophylla 'Dawson's White' 76

Brussels sprouts 242, 255

Buddleja davidii (butterfly bush) 101

    pruning 330

    Buddleja davidii 'Fascinating' 60

budgets 10

bulbs 32, 86–97, 92

    in borders 91, 92

    bulb-like plants for summer and autumn 96–97

    choosing and buying 91

    division 92

    growing 89

    lifting and storing 92, 93

    planting 91, 333, 334

    for spring 34, 90, 94–95

    year-round displays 90

bulblets 336g

butterflies 327

Buxus (box)

    edging 213, 216

    pruning 219

    topiary 29

    Buxus sempervirens 213

        B. s. 'Latifolia Maculata' 221

C

cabbage 255

Calamagrostis 28, 150

    C. × *acutiflora* 'Karl Foerster' 151

calcium 55

Calendula officinalis (pot marigold) 32, 81, 85

Callistephus chinensis Ostrich Plume Series (China aster) 85

Calluna (ling) 75

Caltha palustris (marsh marigold) 53

Camassia leichtlinii (quamash) 95

    C. l. subsp. *leichtlinii* subsp. *leichtlinii* 95

    C. l. 'Semiplena' 53

Camellia japonica 47

    C. j. 'Ace of Hearts' 231

    C. j. 'Guilio Nuccio' 20

Campanula 75

    C. glomerata 'Superba' 21

    C. isophylla 'Stella Blue' 236

    C. poscharskyana 64

Canna (Indian shot plant) 'Endeavour' 97

capsid bugs 316, 319

Carex (sedge)

    C. oshimensis 'Evergold' 38

    C. pendula 52

Carpinus betulus (hornbeam) 58, 213, 219, 220

carrots 248, 250

Caryopteris × *clandonensis* 'Worcester Gold' 21

Catananche caerulea 'Bicolor' (white cupid's dart) 77

caterpillars 318

cauliflower 242, 255

Ceanothus (California lilac)

    C. 'Blue Mound' 100

    C. 'Concha' 112

Centaurea (cornflower/ knapweed) 75, 81, 184

    C. cyanus 57

Cephalaria gigantea (yellow giant scabious) 25

Cercis

    C. canadensis 'Forest Pansy' (Eastern redbud) 206

    C. siliquastrum (Judas tree) 208

Chaenomeles × *superba* 'Crimson and Gold' (flowering quince) 112

chalky soil 55

Chamaecyparis lawsoniana (Lawson cypress) 213

    C. l. 'Pembury Blue' 221

chard 242

cherries 233, 270, 271

chicory 254

Chinese cabbage 255

Chionodoxa 'Pink Giant' (glory of the snow) 94

chitting potatoes 253

Chlorophytum comosum 'Vittatum' (spider plant) 237

Choisya (Mexican orange blossom)

    C. × *dewitteana* 'Aztec Pearl' 100, 116

    C. ternata 'Lich' 231

    C. t. Sundance 38

Chrysanthemum 35

Chusquea culeou (Chilean bamboo) 157, 160

Cistus (sun/rock rose) 107, 283

    C. × *dansereaui* 'Decumbens' 190

    C. salviifolius 43

clay soil 18, 19, 50, 177, 244, 245

Clematis 122

    container-grown 224

    early- and midsummer-flowering 132

    late-flowering 133

    pruning 126

    Clematis alpina 119, 121

    C. armandii 126, 132

    C. 'Bill MacKenzie' 133

    C. 'Blue Ravine' 132

    C. cirrhosa 126

    C. 'étoile Violette' 121

    C. 'Fireworks' 132

    C. 'Hagley Hybrid' 130

    C. 'Huldine' 21, 133

    C. 'Jackmanii' 133

    C. 'Jackmanii Superba' 121

    C. macropetala 126, 132

    C. 'Madame Julia Correvon' 121

    C. montana 126

    C. m. 'Elizabeth' 100

    C. m. var. *grandiflora* 132

    C. 'Niobe' 132

    C. orientalis 75

    C. 'Perle d'Azur' 133

    C. 'Polish Spirit' 133

    C. 'Purpurea Plena Elegans' 133

    C. rehderiana 133

    C. tangutica 75, 133

    C. 'Vyvyan Pennell' 132

    C. 'White Swan' 132

    climbers 118–33

    choosing 121

    fast-growing 131

    for flowers and foliage 128

    how they climb 122

layering 306

planting against wall or fence 124–25, 127, 130

pruning 121, 126–27

    for scent 129

supports 122, 123, 257

where to plant 122–23

see also roses

cloches 336g

clover, white 178

coastal gardens 12, 58, 59, 214

Cobaea scandens (cup and saucer vine) 230

    C. s. f. alba 131

coir 336g

Colchicum 90

    C. speciosum 'Album' 96

cold frames 336g

cold weather, effects of 12–13

colour 6, 30, 36–39

    colour wheel 36, 37

    flower beds and borders 6, 30, 36–39

    grasses 154

common names 340–41

compacted soil 336g

companion planting 243, 325

compost 226, 286, 290–91, 299, 336g, 337g, 338g

composted bark 286, 336g

conifers, watering 282

containers 6, 222–39, 336g

    bamboos in 158

    care and maintenance 228–29

    choosing 225

    feeding 288

    fruit trees 266

    growing crops in 232–33

    planting up 226–27

    plants for 224

    watering 283, 313, 333

Convallaria majalis (lily-of-the-valley) 183

    C. m. 'Albostriata' 48

Convolvulus sabatius 237

coppicing 47, 201, 336g

Cordyline australis (New Zealand cabbage palm) 224, 283

    C. a. 'Albertii' 45

    C. a. 'Variegata' 238

Coreopsis grandiflora 'Badengold' 57

Coriandrum sativum (coriander) 263

corms 89, 91, 336g

Cornus (dogwood) 101, 107
  C. alba 'Sibirica' 114
  C. alternifolia 195
    C. a. 'Argentea' (pagoda
      dogwood/green osier)
      207
  C. controversa 194, 195
  C. kousa 195
    C. k. 'Chinensis' 20
    C. k. 'Miss Satomi' (Japanese
      dogwood) 208
Cortaderia selloana (pampas
  grass) 152
  C. s. 'Silver Comet' 155
Corylus (hazel) 110
  C. avellana 214
    C. a. 'Contorta' (corkscrew
      hazel) 101
Cosmos 36, 75
  C. atrosanguineus (chocolate
    cosmos) 93
  C. bipinnatus Sensation Series 85
  C. b. 'Sonata Pink' 223
Cotinus (smokebush) 101
  C. coggygria Purpureus Group
    56
Cotoneaster 32, 101, 127
  C. horizontalis (fishbone
    cotoneaster) 53, 117
cottage gardens 24
courgettes 259
Crambe maritima (sea kale) 60
Crassula arborescens (silver jade
  plant) 239
Crataegus (hawthorn) 219, 248
  C. laevigata 'Paul's Scarlet' 208
  C. monogyna 213, 214
crocks 336g
Crocosmia (montbretia) 38
  C. masoniorum 44
Crocus 87, 89, 90
  C. minimus 94
  C. ochroleucus (autumn
    crocus) 97
  C. tommasinianus
    'Ruby Giant' 21
crops
  crop rotation 243, 245, 247
  where to grow 242–43
cross-pollination 336g
crowns 337g
  crown lifting of trees 47
Cryptogramma crispa (parsley
  fern) 163
cucumbers 257, 259

cultivars 337g, 341
× *Cupressocyparis leylandii*
  (Leyland cypress) 214
currants 264, 265, 272, 275
  see also blackcurrants;
    redcurrants; whitecurrants
cut flowers, perennials for 79
cuttings 7, 308–11, 337g
Cycas revoluta (Japanese sago
  palm) 239
Cyclamen 38, 89, 183
  C. hederifolium 97
cypress, swamp 51
Cytisus (broom) × *praecox*
  'Allgold' 45

D

Dahlia 89, 92, 93
  D. 'Arabian Night' 96
  D. 'David Howard' 79
  D. 'Yellow Hammer' 230
daisy 179, 341
damping off 323, 337g
dandelion 178, 299
Daphne 101
  D. bholua 114
  D. mezereum 116
daylily 34
deadheading 75
deciduous 337g
Delphinium 35, 71, 74, 75
  D. 'Fenella' 79
Deschampsia cespitosa
  'Goldtau' (tufted hair grass)
  155, 190
designing your garden 26–27
Deutzia 108
Dianthus (pink)
  D. alpinus (alpine pink) 63
  D. 'Little Jock' 191
Diascia barberae 'Blackthorn
  Apricot' 236
Dicksonia antarctica (tree fern)
  163
Dierama pulcherrimum (angel's
  fishing rod) 190
digging 244, 279, 292–93, 337g
Digitalis (foxglove) 32, 41, 74, 75
  D. purpurea 53
diseases 245, 247, 267, 320–33
disorders 314
division 304–05, 337g
dormancy 337g
doronicum 304

double-digging 337g
drainage 13, 43, 50, 51
drills 336g
drip-irrigation systems 284
drought 51, 314
  lawns and 176
drought-tolerant plants 43, 55,
  56–57, 283, 285
dry gardens 42, 54–57, 297
Dryopteris
  D. affinis (golden male fern)
    164
  D. filix-mas (male fern) 164

E

earwigs 316
Eccremocarpus scaber (Chilean
  glory flower) 128
Echeveria pulvinata (plush plant)
  239
Echinops (globe thistle) 28, 41,
  283
  Echinops ritro 41
    E. r. 'Veitch's Blue' 191
eco-gardening 296–97
eelworms 318
Elymus magellanicus (Magellan
  rye grass) 154
endives 233
Ensete ventricosum (Abyssinian/
  Ethiopian banana) 239
entertaining 10
Epimedium (barrenwort/
  bishop's mitre)
  E. grandiflorum 'Crimson
    Beauty' 78
  E. × *perralchicum* 48, 75
Eranthis hyemalis (winter
  aconite) 90, 92, 95
Erica (heath)
  E. carnea 75
    E. c. 'Vivellii' 115
  E. vagans 'Birch Glow' (Cornish
    heath) 20
ericaceous compost 337g
ericaceous plants 19
Eriobotrya japonica (loquat)
  206
Eryngium (sea holly) 32, 75, 283
  E. giganteum 187, 191
  E. × *oliverianum* 55
  E. × *tripartitum* 44
Erythronium 'Pagoda'
  (dog's-tooth violet) 95

Escallonia 101
  pruning 219
  Escallonia 'Langleyensis' 117
Eschscholzia californica
  (California poppy) 84
Eucomis 90
Euonymus 100
  E. europaeus (spindle tree)
    214
    E. e. 'Red Cascade' 113
  E. fortunei (evergreen
    bittersweet)
    E. f. 'Emerald Gaiety' 117
    E. f. 'Emerald 'n' Gold' 114
    E. f. 'Silver Queen' 38
Eupatorium purpureum (Joe
  Pye weed) 52
Euphorbia (milkweed/spurge)
  characias 31, 115, 187
exotic plants, for containers
  238–39
exposed sites 58–61, 213

F

F1 hybrid 337g
Fagus sylvatica (beech) 213, 219,
  221
Fallopia baldschuanica
  (mile-a-minute plant) 131
families, plant 341
family gardens 10
Fargesia 156, 157
  F. murielae (umbrella/Muriel's
    bamboo) 156, 157, 161
    F. m. 'Simba' 161
  F. nitida (fountain bamboo)
    157, 160
Fatsia japonica (Japanese aralia)
  38, 100, 115, 231
feeding plants 288–89
  fruit bushes and trees 267
Felicia amelloides 'Read's White'
  236
ferns 149, 162–65, 213, 224
  hardy ferns 164
  see also individual types of
    fern
fertilization 337g
fertilizer 247, 267, 337g
  controlled-release 229
  water-soluble 229
Festuca glauca (blue fescue)
  154
  F. g. 'Elijah Blue' 38, 61

Ficus carica (fig) 233
    F. c. 'Brown Turkey' 206
flax, New Zealand 31
fleece 337g
flowers, grasses for 155
Foeniculum vulgare (fennel) 262
Forsythia 101
    pruning 108, 219
    Forsythia suspensa 21, 117
Fothergilla gardenii (witch alder) 20
Fremontodendron 'California Glory' (flannel flower) 117
French beans 256
Fritillaria
    F. camschatcensis (black sarana/chocolate lily) 97
    F. meleagris (snake's head fritillary) 94
frost 12–13, 15, 264, 315
    frost pockets 17, 337g
    ground frost 12
fruit 233, 283, 288, 323
    fruit bushes 264, 265, 267, 272–73
    fruit trees 264, 266–67, 268–69
    growing 264–65, 266
    pruning and caring for 266–67
    soft fruits 272–73
    where to grow 242–43
    see also individual types of fruit
Fuchsia 101
    F. 'Andrew Hadfield' 113
    F. 'Corallina' 237
    F. 'Mrs Popple' 49
    F. 'Thalia' 231
fungi 320
fungicides 321

G

Gaillardia × grandiflora 'Dazzler' (blanket flower) 79
Galanthus (snowdrop) 38, 47, 90, 92, 183
    G. 'Atkinsii' 95
    G. nivalis Sandersii Group 52
galls 318
Galtonia viridiflora 97
garden centres 66
garlic 251
Garrya elliptica 117

Gaultheria tasmanica (pernettya) 20
gel, water-retaining 229, 285
Genista hispanica (Spanish broom) 43
Gentiana septemfida 64
genus 341
geotextile see membranes
Geranium (cranesbill) 75, 187
    G. cinereum 'Ballerina' 65
    G. phaeum 49
    G. tuberosum 55
germination 337g
Ginkgo biloba (maidenhair tree) 209
Gladiolus 89, 90, 92, 93
    G. tristis 94
Glandularia 'Diamond Merci' 236
Glaucium flavum (yellow-horned poppy) 61
glaucous 337g
Glechoma hederacea 'Variegata' (ground ivy) 237
gooseberries 264
grasses 31, 37, 148–55, 283
    annual 84
    autumn/winter gardens 28
    for colour 41
    planting 151, 152–53
    seedheads 75
gravel 153, 183, 286, 287
gravel gardens 184–91, 297
    planning 186
    plants for 187, 190–91
    preparing planting site 187
green manure 337g
greenfly 316
Griselinia littoralis 58, 60, 213
ground cover 71, 183
ground frost 12
grow bags 246
growing seasons 337g
Gypsophila (baby's breath)
    G. muralis 'Gypsy Pink' 223
    G. repens 'Dorothy Teacher' (baby's breath) 65

H

Hakonechloa macra 'Aureola' (hakone grass) 154
half-hardy plants 13, 337g
Hamamelis (witch hazel) 101
    H. × intermedia 'Jelena' 114

hanging baskets 224, 233, 234–35, 236–37, 283, 288, 332
hard landscaping 337g
hardening off 303, 337g
hardiness 13, 337g
hardwood cuttings 308
hardy plants 13, 337g
heavy soil 337g
Hebe
    H. 'Bowles' Variety' 57
    H. 'Great Orme' 21
    H. ochracea 'James Stirling' 38
    H. pinguifolia 'Pagei' 38
Hedera (ivy) 120, 122
    layering 306
    variegated 339g
    Hedera helix
        H. h 'Dragon Claw' 130
        H. h. 'Eva' 128
        H. h. 'Glymii' 48
        H. h. 'Little Diamond' 236
    H. hibernica 38, 130
    H. pastuchovii subsp. cypria 130
hedge trimmers 219
hedges 58, 210–21
    buying plants 215
    clipping and shaping 218, 219, 332, 333
    competing for water 71
    planting 2167
    renovating 213, 219
    for wildlife 214
heeling in 337g
Helenium (Helen's flower) 38
    H. 'Crimson Beauty' 79
Helianthemum (rock rose/sun rose) 283
    H. nummularium 43
    H. 'Wisley Primrose' 113
Helianthus (sunflower)
    H. annuus 341
        H. a. 'Teddy Bear' 341
    H. decapetalus 341
    H. × multiflorus 341
    H. 'Triomphe de Gand' 84
Helichrysum
    H. italicum (curry plant) 60
    H. petiolare 'Limelight' 237
Helictotrichon sempervirens (blue oat grass) 154

Helleborus 183
    H. orientalis 49
    H. torquatus 'Dido' 76
Hepatica nobilis var. japonica 76
herbaceous plants 29, 337g
herbicides 337g
herbs 43, 54, 183, 233, 261, 262–63
    in containers 233
    growing 261
    in paving 183
Hertia cheirifolia (othonna) 60
Hibiscus syriacus 'Lady Stanley' 45
hill gardens 13
Hippophae rhamnoides (sea buckthorn) 61, 213
hoes 279, 293
Holcus
    H. lanatus (Yorkshire fog) 179
    H. mollis 'Albovariegatus' (striped Yorkshire fog) 154
hollyhock rust 322
honey fungus 322
hops 287
horticultural grit/sand 337g
Hosta 31, 37, 224
    division 305
    Hosta 'Buckshaw Blue' 48
    H. 'Francee' 78
Humulus lupulus 'Aureus' (golden hop) 131
humus 337g
Hyacinthoides non-scripta (bluebell) 48
Hyacinthus orientalis 90
    H. o. 'Blue Jacket' 230
    H. o. 'Ostara' 94
hybrids 341
Hyde Hall 42, 59
Hydrangea 101, 106
    H. anomola subsp. petiolaris (climbing hydrangea) 120, 130
    H. macrophylla 106
        H. m. 'Maculata' 113
        H. m. 'Nikko Blue' 231
    H. paniculata 'Praecox' 48
        H. p. 'Unique' 99
hygiene 320, 324–27
Hylotelephium spectabile 'Brilliant' 191
Hypericum (St John's wort) 101
    H. × hidcoteense 75

I

Ilex (holly) 213
 pruning 218, 219
 Ilex aquifolium 214
  I. a. 'Golden Milkboy' 38,
   220
  I. a. 'Madame Briot' 114
insecticides 317
inspiration, garden 225
intercropping 242
Inula hookeri 77
Ipheion 90
Ipomoea (morning glory) 36,
 120, 121
 I. coccinea 131
 I. purpurea 85
Iris 22, 89, 90, 332, 338
 division 305
 Iris 'George' 94
 I. latifolia 96
 I. unguicularis 'Mary Barnard'
  190

J

Jasminum (jasmine) 224
 J. nudiflorum 117
 J. officinale
  'Argenteovariegatum' 129
John Innes composts 337g
Juniperus communis
 'Compressa' 65
 J. squamata 'Blue Star' 38
 J. s. 'Meyeri' 60

K

kale 255
Kalmia latifolia (calico bush)
 113
Kerria (Jew's mantle) 100
 pruning 108
Kerria japonica (Japanese rose)
 'Golden Guinea' 112
knapweed see Centaurea
Knautia macedonica 25
Kniphofia triangularis (red-hot
 poker/torch lily) 77
knot gardens 43
Koelreuteria paniculata (golden
 rain tree/pride of India) 207
kohl rabi 250
Kolkwitzia amabilis 'Pink Cloud'
 (beauty bush) 112

L

Laburnum × watereri 'Vossii' 208
Lagurus ovatus (hare's tail) 84
Lamium maculatum (dead
 nettle) 'Album' 49
Lamprocapnos spectabilis
 (bleeding heart) 49
landscape fabric see membranes
landscaping, hard 337g
Lantana camara 'Radiation' 238
lateral 338g
Lathyrus (sweet peas) 75, 120,
 121, 122
 L. latifolius (everlasting pea)
  130
 L. odoratus (sweet pea) 84,
  129
Latin names 340–41
Laurus nobilis (bay) 262
 L. n. f. angustifolia (sweet bay/
  bay laurel) 115
Lavandula (lavender) 216, 283
 hedging 216
 Lavandula angustifolia 187
  L. a. 'Hidcote' 220
  L. a. 'Munstead' 56
  L. a. 'Twickel Purple' 190
  L. stoechas 115
lawns 166–83
 autumn lawn care 172–73
 edge repair 174–75
 feeding 171, 288
 from seed 182
 low-maintenance alternatives
  183
 moss 177, 183
 mowing 170–71
 planting bulbs in 92
 sandy soil 55
 size and shape 169
 troubleshooting 176–77
 turf 167, 180–81
 watering 171, 283
 waterlogging 176–77
 weeds 177, 178–79
layering 219, 306–307, 338g
leader/leading shoot 338g
leaf miners 319
leaf mould 286, 287, 338g
leafy vegetables 254–55
leaves
 for colour 101
 pest and disease damage 102,
  318–19, 322–23
legumes 338g

lettuce 242, 254
Leucanthemum vulgare
 (ox-eye daisy) 24, 184
Leucothoe fontanesiana
 (switch ivy) 'Rainbow' 114
Leucothoe 'Scarletta' 112
Leycesteria formosa
 (Himalayan honeysuckle) 113
Leymus arenarius 38
light soil 338g
Ligularia dentata 'Desdemona'
 77
Ligustrum ovalifolium (privet)
 212, 213, 218, 219, 221
 pruning 218, 219
 Ligustrum ovalifolium 'Vicaryi'
  100
Lilium (lily) 90, 340
 in containers 224
 Lilium henryii 96
 L. regale (Regal lily) 96
lily beetle 316
lime 245, 246, 338g
limes 233
Limnanthes douglasii
 (poached egg plant) 64
limy soil see alkaline soil
Liquidambar orientalis
 (oriental sweet gum) 209
Liriodendron tulipifera
 'Aureomarginata' 195
Liriope muscari (lilyturf) 77, 183
loam 18, 338g
Lobelia 75
 L. 'Colour Cascade' 236
Lonicera (honeysuckle) 24,
 122
 layering 306
 pruning 219
 Lonicera × americana 130
 L. nitida (shrubby
  honeysuckle) 213, 216
  L. n. 'Baggesen's Gold' 220
 L. periclymenum 'Serotina'
  129
Lonicera nitida (box) 107
Lotus berthelotii (parrot's beak/
 coral gem) 236
low-maintenance gardens,
 plants for 35
Lupinus (lupin) 25, 75
Luzula
 L. campestris (field woodrush)
  179
 L. nivea (woodrush) 49

Lychnis chalcedonica
 (Jerusalem/Maltese cross)
 79
Lysimachia ciliata 'Purpurea'
 (loosestrife) 52

M

magnesium 55
Magnolia 101
 M. denudata (lily tree) 207
 M. × soulangeana 'Lennei Alba'
  208
Mahonia
 pruning 107
 Mahonia × media 'Charity'
  114
Malus (apple) 101, 233, 264, 268
 M. × arnoldiana (crab apple)
  208
 M. 'Profusion' 195
 M. × purpurea 'Lemoinei'
  (crab apple) 21
 M. 'Red Devil' 208
manure 245, 247, 286, 337g
marginals 338g
marrows 259
Matteuccia struthiopteris
 (ostrich/shuttlecock
 fern) 164
mature gardens 11
meadow gardens 184–85
 annual 184, 185
Melissa officinalis 'Aurea' (lemon
 balm) 262
membranes 187, 188–89, 287,
 338g
Mentha spp. (mint) 262
Michaelmas daisy 35
microclimate 16–17, 338g
mildew 320
millipedes 327
mind-your-own-business 179
mint 233
Miscanthus sinensis 28, 150, 153
 pruning 153
 Miscanthus sinensis 'Grosse
  Fontäne' 155
  M. s. 'Kleine Silberspinne' 151
  M. s. 'Variegatus' 154
  M. s. 'Zebrinus' (zebra grass)
   154
mixed borders 22, 27, 29, 69,
 145, 338g
modules 298, 338g

moles 177, 325
Molinia caerulea 28, 150
   M. c. subsp. *arundinacea*
      'Transparent' 151
   M. c. subsp. *a.* 'Windspiel' 151
   M. c. subsp. *caerulea*
      'Variegata' (striped purple
      moor grass) 154
Monarda 'Mahogany'
   (bergamot) 53
monochrome planting 36–37
mosaic virus 322
moss in lawns 177, 178, 183
mould 320, 323
mowers 170–71
mulching 18, 272, 285, 286–87,
   293, 338g, 339g
   trees 200
Muscari armeniacum
   (grape hyacinth) 94
mushroom compost 287, 338g

N

Nandina domestica
   (heavenly bamboo) 56
Narcissus (daffodil) 32, 38, 89
   N. 'February Gold' 90
   N. 'Satin Pink' 95
   N. 'Suzy' 94
   N. 'Tête-à-Tête' 65
   pheasant's eye 90
native plants 338g
naturalistic 338g
naturalized 338g
nectarines 271
neglected gardens 11
neighbours and hedges 214
Nemesia caerulea 237
Nepeta (catmint) 75, 186
   N. sibirica 35
Nephrolepis exaltata
   (sword fern) 165
Nerine 90
   N. bowdenii 97
neutral soil 339g
new gardens 10, 11
New Zealand flax 31
Nicotiana 'Lime Green'
   (tobacco plant) 85
Nigella damascena
   (love-in-a-mist) 81, 84
nitrogen 289, 315
notching 267
NPK 338g

nurseries 66
nursery bed 338g
nutrients 289, 338g
   lack of nutrients 315
Nyssa sinensis (Chinese tupelo)
   209

O

Ocimum basilicum (basil) 262
offsets 338g
Olearia 213
onions 250, 251
   see also Allium
Ophiopogon planiscapus
   (lilyturf)
   O. p. 'Kokuryu' 38
   O. p. 'Nigrescens' 78
organic gardening 296
organic matter 245, 247, 265, 338g
   digging in 9, 292, 330
   for dry soils 55, 285
   as mulches 286–87
   for perennials 72
   for shady sites 47
   for wet soils 51, 177
Origanum (marjoram/oregano)
   O. amanum 63
   O. 'Kent Beauty' 115
   O. onites 263
Ornithogalum dubium
   (star-of-Bethlehem) 95
Osmanthus 101
   O. × *burkwoodii* 116
Osmunda regalis (royal fern) 164
Osteospermum 'Buttermilk' 236
overwintering 81, 338g

P

Pachysandra terminalis 78
Paeonia
   P. delavayi (tree peony) 100,
      112
   P. officinalis 'Rubra Plena' 79
pak choi 255
Panicum virgatum 'Dallas Blues'
   (blue switch grass) 155
Papaver (poppy) 75
   seeds 298
   Papaver orientale 187
   P. o. 'Patty's Plum' 22
   P. rhoeas (field poppy) 184
   P. r. Mother of Pearl Group
      44

P. somniferum (opium poppy)
   84
Parrotia persica (Persian
   ironwood) 209
parsnips 248, 250
parterres 43
Parthenocissus (Virginia
   creeper) 122, 126
   P. henryana (Chinese Virginia
      creeper) 131
   P. tricuspidata (Boston ivy) 131
Passiflora (passion flower) 123
   P. caerulea 131
   P. c. 'Constance Elliott' 128
paths 261
   between beds 31
   gravel 43
   mown grass 24, 169
patios, roses for 147
paving, herbs in 183, 187
pea sticks 338g
peaches 233, 271
pears 233, 269
pearlwort 178
peas 246, 256, 257
peat substitution 338g
Pelargonium 75
   P. 'Apple Blossom Rosebud' 44
   P. 'Lemon Fancy' 230
   P. 'Mr Henry Cox' 238
Pennisetum (fountain grass)
   P. alopecuroides 57
   P. a. 'Hamelin' 155, 190
   P. setaceum 'Purpureum' 155
Penstemon 'Apple Blossom' 56
peppers 233, 246, 259
perennials 69, 70–79, 338g
   for autumn interest 77
   for containers 230
   for cut flowers 79
   deadheading 75
   designing with 32
   division 304–05
   planting 72–73
   for shade 78
   supports 74
   watering 283
   for winter and spring interest
      76
pergolas 123, 144, 338g
   roses for 144
Pericallis × *hybrida* (cineraria)
   238
Persicaria 329
   P. affinis 'Donald Lowndes' 53

pesticides 317, 321, 338g
pests 243, 245, 247, 267, 313,
   316–19, 321
   traps and barriers 325
Petroselinum crispum (parsley)
   263
pH of soil 19, 47, 245, 247, 338g
Phalaris arundinacea (gardener's
   garters) 151
Philadelphus (mock orange) 101
   P. coronarius 'Variegatus' 15
   P. 'Virginal' 116
Phlomis 283
   P. cashmeriana 44
   P. fruticosa (Jerusalem sage)
      113
   P. russeliana 77
Phormium 283
   P. tenax Purpureum Group 38
phosphorus 289
Photinia
   P. davidiana 21
   P. × *fraseri* 'Red Robin' 112
photosynthesis 338g
Phyllostachys (bamboo) 156, 157
   P. aureosulcata f. *aureocaulis*
      (golden/fishpole bamboo)
      160
      P. a. f. *spectabilis* 149
   P. bambusoides 'Castilloni' 157
   P. nigra (black bamboo) 157,
      160
      P. n. henonis (Henonis
      bamboo) 161
   P. viridiglaucescens 160
   P. vivax f. *aureocaulis* (golden
      Chinese timber bamboo) 160
Physalis alkekengi (Chinese/
   Japanese lantern) 35, 77
Physocarpus 261
pinching out 338g
Pittosporum tobira 38, 115
The Plant Finder 66
Plantago major 179
plantain, broad-leaved 179
plants (general only)
   for acid soil 20
   for alkaline soil 21
   choosing and buying 32–39,
      66–67
   for containers 224, 230–31
   for different seasons 15,
      34–35, 38, 90, 101
   drought-tolerant 43, 55,
      56–57, 283

for exposed gardens 59, 60–61
feeding 288–89
for gravel garden 187, 190–91
for hanging baskets 236–37
hardiness 13
for hedges 220–21
for low-maintenance
    garden 35
planting plan 40–41
reproduction 299
for rock gardens 62
for shade 47, 48–49
shape and foliage 41
for structure 100
for sunny sites 43, 44–45
tolerating wet soil 51–53
understanding names 340–41
see also individual plants
plastic sheeting, black 287
Platycerium bifurcatum
    (common staghorn
    fern) 165
Pleioblastus 156
    P. pygmaeus 158
    P. variegatus (dwarf white
        stripe bamboo) 157, 161
plug plants 302
Plumbago auriculata (Cape
    leadwort) 230
plums 270
Poa annua 178
pollarding 47, 201, 338g
pollen 299
pollen beetles 317
pollination 336g, 338g
pollinators 338g
Polypodium vulgare
    (common polypody) 164
Polystichum setiferum
    (soft shield fern) 47, 164
ponds 326
pot bound 338g
potassium 289
potatoes 250, 253
Potentilla fruticosa (shrubby
    cinquefoil)
    P. f. Princess 113
    P. f. 'Red Ace' 61
potting on 338g
powdery mildews 323
pricking out 303, 338g
Primula
    P. dentata 304
    P. veris (cowslip) 76
propagation 253, 298–311, 339g

propagators 246, 299, 339g
Prunella vulgaris 178
pruning 330, 332
    climbers 121, 126–27
    fruit bushes 265
    fruit trees 266–67
    hedges 211, 218
    roses 140–43
    shrubs 103, 106–11, 335
    trees 47, 200–01, 202–05
Prunus (cherry)
    pruning 218, 219
    Prunus cerasifera 'Nigra' 195
    P. laurocerasus (cherry laurel)
        218
    P. lusitanica (Portugal laurel)
        220
    P. serrula 195
    P. spinosa 'Purpurea'
        (blackthorn/sloe) 60
    P. 'Spire' 208
Pseudosasa japonica (arrow
    bamboo) 156, 160
Pteris cretica var. *albolineata*
    (Cretan brake) 165
Pulmonaria (lungwort)
    division 304
    Pulmonaria 'Mawson's Blue' 78
    P. saccharata (Bethlehem
        sage) 48, 75
Pulsatilla vulgaris 55
pumpkins 260
purple plants 38
Pyracantha (firethorn)
    pruning 219
    training 127
    Pyracantha 'Watereri' 117
Pyrus salicifolia 'Pendula'
    (weeping pear) 195, 207

R

radishes 248
rain shadow 13, 71, 124, 282,
    339g
raised beds 43, 50, 232, 244,
    292
raking, lawns 172–73
Ranunculus repens 178
raspberries 265, 273
redcurrants 264, 265, 274
repotting containers 229
reverted 339g
rhizomes 89, 339g
Rhodiola rosea (rose root) 43

Rhodochiton atrosanguineus
    128
Rhododendron 336
    R. 'Irohayama' 112, 231
    R. 'Palestrina' 20
Rhus typhina (stag's horn
    sumach/velvet sumach) 209
Ribes sanguineum (flowering
    currant) 101
ripening 339g
Robinia pseudoacacia 'Frisia'
    (false acacia/black locust)
    207
rock gardens 62–65
    creating 62
    plants for 63, 64–65
rocket 254
Rodgersia 51
    R. pinnata 'Superba' 53
root cuttings 308, 309
root vegetables 248–53
roots 67, 339g
    checking for health 102
    disorders 315
    rootballs 339g
    rootstocks 339g
Rosa spp.
    R. 'Alba Maxima' 139, 168
    R. 'Albertine' 122, 144
    R. Anna Ford 147
    R. 'Arthur Bell' 146
    R. Baby Love 145
    R. Baby Masquerade 147
    R. 'Ballerina' 145
    R. Baronne Edmond de
        Rothschild 146
    R. 'Bobbie James' 122
    R. Bonica 139, 145
    R. Breath of Life 146
    R. 'Buff Beauty' 135
    R. canina (dog rose) 139
    R. 'Charles de Mills' 145
    R. Cider Cup 147
    R. 'Climbing Iceberg' 137, 144
    R. 'Compassion' 144
    R. 'Complicata' 145
    R. 'Constance Spry' 145
    R. 'Félicité Parmentier' 146
    R. Ferdy 147
    R. filipes 'Kiftsgate' 139
    R. Flower Carpet series 139
    R. Fragrant Cloud 146
    R. 'Geranium' 138
    R. Gertrude Jekyll 146
    R. glauca 138, 139

R. 'Great Maiden's Blush' 139
R. Handel 144
R. 'Hertfordshire' 59, 60
R. L'Aimant 52
R. Laura Ashley 147
R. Laura Ford 144
R. Little Bo-Peep 147
R. 'Madame Hardy' 145
R. 'Marguerite Hilling' 138, 139
R. moyesii 139
R. 'Nathalie Nypels' 147
R. 'Nevada' 138
R. 'New Dawn' 144
R. 'Penelope' 146
R. 'Perle d'Azur' 122
R. 'Phyllis Bide' 144
R. pimpinellifolia 139
R. 'Precious Platinum' 137
R. primula 138
R. 'The Queen Elizabeth' 137
R. 'Roseraie de l'Haÿ' 145
R. rubiginosa 138
R. rugosa (hedgehog rose) 221
    R. r. 'Alba' 139
    R. r. 'Fru Dagmar Hastrup'
        139
    R. r. 'Nyveldt's White' 139
R. 'Sally Holmes' 137
R. 'Sander's White Rambler'
    137
R. Summer Wine 144
R. Valencia 146
R. Warm Welcome 147
R. xanthina f. *hugonis* 138
Roses 122, 134–47
    for berries 101
    choosing and buying 138–39
    climbing 137
    damask 141
    deadheading 139
    family 340
    feeding 288
    floribunda 137, 140
    gallica 141
    growing 136–39
    hybrid tea 137, 140
    for mixed beds 145
    moss roses 141
    for patios 147
    for pergolas and arches 144
    planting 138
    pruning 140–43
    rambling 122, 137, 142
    repeat-flowering 75
    for scent 146

shrub roses 135, 137, 139
siting 139
species 138, 139
Roscoea cautleyoides 96
rosehips 139
roses *see Rosa*
Rubus (bramble) 122
pruning 107
Rudbeckia 31, 34
Rumex acetosa (sorrel) 263
runner beans 233, 256, 257
runners 339g
rusts 322
Ruta graveolens 'Jackman's Blue' 38

S

sage 29, 48, 75, 261
Sagina procumbens 178
salad leaves 254–55
Salix (willow)
pruning 107, 110
Salix caprea 'Kilmarnock' (Kilmarnock willow) 231
Salvia 25, 75
S. 'Blue Spire' 57
S. *officinalis* 263
S. *o.* 'Purpurascens' 38
S. *o.* 'Tricolor' 191
S. *patens* 45
S. *p.* 'Cambridge Blue' 77
Salvia rosmarinus (rosemary) 61, 263, 283
Sambucus (elder)
pruning 110
Sambucus racemosa 'Sutherland Gold' 100
sandy soil 18, 19, 54, 55, 151, 244
Sansevieria trifasciata 'Golden Hahnii' (mother-in-law's tongue) 239
Santolina (lavender cotton) 283
S. *chamaecyparissus* 44
Sarcococca confusa (Christmas box/winter box) 101, 116
Sasa 156, 157
S. *palmata* f. *nebulosa* 160
S. *veitchii* (Veitch's bamboo) 161
saw, pruning 278
sawfly, leaf rolling 318
Saxifraga sancta 65
Scabiosa lucida (scabious) 56
Scaevola aemula (fairy fan flower) 237

scale insects 316
scarification 339g
scent 116
climbers for 129
roses for 146
shrubs for 116
Schizophragma integrifolium 130
Scilla peruviana 'Alba' (squill) 97
screens, bamboo 157
seasonal effects 15, 31
seating areas 23, 24, 25
secateurs 279, 281
Sedum (stonecrop) 28, 41, 75
S. *spathulifolium* 'Cape Blanco' 64
S. *spectabile* 187
seed leaves 339g
seedbeds, preparing 245
seedheads 28, 75
seedlings 298–9, 302–03, 331
seeds 298–301
annuals and biennials 81
collecting 81, 300
for lawns 172–73, 176, 180, 182
sowing 82–83, 246, 300–01
storing 300
watering 282
wildflower meadows 185
seep hoses 284, 339g
self-heal 178
self-seeding 339g
semi-ripe cuttings 308, 310–11
Semiarundinaria fastuosa var. *viridis* (Narihara bamboo) 161
Sempervivum ciliosum (houseleek) 56
serpentine layering 306
shade 15, 46–49, 324
causing disorders 315
damp 17
perennials for 78
plants for 47, 48–49
trees and 47, 194
shallots 251
sheet mulch 339g
shelter belts 17, 58, 59, 339g
shrubs 32, 98–117
choosing and buying 102
for containers 231
evergreen 115
feeding 288
for fragrance 116
planting 103
pruning 103, 106–11, 335
for spring colour 112

for summer and autumn colour 113
transplanting 1045
for walls 117, 339g
whips 59, 339g
for winter colour 114
for year-round interest 101
see also individual plants
Silene schafta (campion/ catchfly) 64
silver leaf disease 323
silver plants 42, 43
single digging 294
Sisyrinchium 25
S. *striatum* 'Aunt May' 190
Skimmia japonica 'Bronze Knight' 20, 114
slugs and snails 316, 318, 319, 325, 327
snow 335
soft fruits 272–73
softwood cuttings 308, 309
soil 18–19
assessment of 11
compacted 336g
dry and free-draining 54–57
improving 46, 47, 51
and moisture 13
neutral 338g
pH of 19, 47, 245, 247, 338g
preparing 180, 244–45
types 18–19
wet and sticky 50–53
soil mark 339g
Solandra maxima (chalice vine/ cup of gold) 129
Solanum crispum 'Glasnevin' 128
Soleirolia soleirolii 179
sooty mould 319
Sorbus (mountain ash/rowan) 31
S. *aria* 'Lutescens' (whitebeam) 206
S. *commixta* (Japanese mountain ash/rowan) 209
S. *sargentiana* (Sargent rowan) 209
sour 339g
sowing vegetables 246
spades 279, 280
species 341
specimen shrubs 100, 339g
speedwell, slender 178
spinach 254
spit 339g

spring 15
bedding plants 34
bulbs for 34, 90, 94–95
colour in 38
garden tasks 330–31
lawn feed 171
perennials for 76
shade plants for 47
shrubs for 101, 112
trees for blossom 208
spring onions 233, 251
sprinklers 171, 284
sprouting broccoli 255
squashes
summer 257, 260
winter 260
squirrels 325
stakes *see* supports
standards 201
stem cuttings 308
Stewartia monadelpha 207
Stipa
S. *arundinacea* (pheasant's eye grass) 44
S. *gigantea* (giant feather grass/golden oats) 28, 38, 150, 155
storing root vegetables 250
straw, composted 287
strawberries 264, 272
in containers 233
stress 339g
structural plants 40, 100
stunted growth 58
subsoil 18, 339g
succulents 339g
suckers 201, 339g
summer
bulb-like plants for 96–97
garden tasks 332–33
shrubs for 101, 113
summer colour 38, 113
summer flowers 35
summer squashes 257, 260
sun scorch 315
sunny sites 14, 17, 42–45
climbers for 120
supports 331, 338g
for climbers 122, 123
for perennials 74
for trees 196
for vegetables 257
swamp cypress 51
Swiss chard 242
Syringa 101

T

Tamarix (tamarisk) 213
  T. ramosissima 61
tamp 339g
tap roots 339g
Taraxacum officinale 178
Taxus (yew)
  pruning 107, 219
  Taxus baccata 213, 221
    T. b. 'Fastigiata' (Irish yew) 207
tender plants 13, 339g
tendrils 339g
Thamnocalamus spathiflorus 161
thatch 339g
thinning 266
Thunbergia alata (black-eyed
  Susan) 121, 230
Thymus (thyme) 25, 100, 187,
  283
  T. 'Bressingham' 64
  T. pulegioides 'Aureus'
    (lemon thyme) 263
Tiarella cordifolia (foam flower)
  49
Tigridia pavonia (tiger flower) 96
tilth 339g
tip layering 306
tomatoes 242, 246, 258
tools 276–79, 280–81
  pruning 278
  storing and caring for 280–81
top-dressing 339g
  containers 229
  lawn 172–73
topsoil 18, 180, 339g
  adding grit 43
Trachelospermum 123
  T. jasminoides (star jasmine)
    116, 129
Trachycarpus fortunei
  (Chusan palm) 231
transpiration 339g
trees 192–209
  for autumn interest 209
  caring for 200–01
  climbers up 122, 123
  competing for water 71
  for containers 231, 233
  creating shade 47, 194
  for focal points 207
  fruit trees 264, 266–67
  planting 196–97, 198–99
  positioning 194–95
  pruning 200–01, 202–05
  for small gardens 206

for spring blossom 208
stunted 58
supports 196
watering 200
trefoil, lesser yellow 179
trellis 123, 257
Trifolium
  T. dubium (lesser yellow
    trefoil) 179
  T. repens (white clover) 178
Trillium 90
  T. grandiflorum 48
Tropaeolum (nasturtium) 121
  climbing 120
  T. majus 85
  T. peregrinum (canary
    creeper) 230
  T. speciosum (flame
    nasturtium) 128
true plants 339g
tubers 89
  planting 91
Tulipa (tulips) 55, 89, 90
  in containers 224
  lifting 92
  T. 'Ancilla' 45
  T. 'Ballerina' 95
  T. 'Rococo' 90
  T. 'Showwinner' 87
  T. sylvestris 90
turf see lawns

V

valley gardens 12
variegated plants 15, 339g
variegation 339g
variety 339g
vegetables 23, 232–33, 283,
  288
  growing 242–43, 244–60
  preparing the soil 244–45
  sowing 246
Verbascum (mullein) 22, 74,
  283
  V. 'Cotswold Queen' 57
Verbena 75
  V. bonariensis 31, 74
vermiculite 299, 339g
Veronica filiformis 178
Viburnum
  V. × bodnantense 106
  V. × burkwoodii 'Anne Russell'
    116
  V. davidii 100

V. plicatum f. tomentosum
  'Mariesii' 100, 101
V. tinus 'Eve Price' 49
  V. t. Variegatum' 115
Vinca (periwinkle) 75
  V. minor 183
vine weevils 317, 319, 327
Viola (pansy) 35
  V. 'Jackanapes' 237
  V. wittrockiana Crystal Bowl
    Series 65
Virginia creeper see
  × Parthenocissus
viruses 321
Vitis (vine) 123
  V. coignetiae 128

W

walls
  climbers for 123, 124–25, 127,
    130
  containers on 224
  planting for 17
  rain shadow 71
  shrubs for 117, 339g
  sunny 14, 17
wasps 327
water butts 285, 297
water shoots 201
water usage reduction 285,
  297
watering 282–85
  containers 228, 283, 333
  eco-gardening 297
  fruit trees and bushes 267
  lawns 171, 283
  overwatering 314, 326
  trees 200
  underwatering 313
waterlogging, lawns 176–77
weather 12–17
weeds and weeding 71, 285, 293,
  326, 332
  fruit trees and bushes 267
  gravel gardens 187
  lawns 177, 178–79
  neglected gardens 11
  weed-supressing membranes
    339g
Weigela 'Looymansii Aurea' 57
wet sites 50–53
whips 59, 201, 339g
white plants 38
whitecurrants 265, 274, 275, 364

whitefly 326, 327
wild gardens 11
wildflower meadows 184–85
wildlife
  encouraging 327
  hedges for 214
  lawns and 169
  wildlife gardens 10
wind
  damage 58, 74
  turbulence 16–17, 213
  windbreaks 58–59, 211, 285
winter 7, 15
  biennials 81
  colour in 38
  garden tasks 334
  grasses 151
  herbaceous plants 28
  low-maintenance garden 35
  perennials for 76
  shrubs for 101, 114
  tree pruning 202–05
  winter flowers 35
Wisteria 121, 122, 123
  pruning 126, 127
  Wisteria floribunda 'Alba' 129
wood chips 287
woodland gardens 324
woodlice 327
woodrush, field 179

Y

yarrow 179
year planner 328–35
yellow-horned poppy 61
yellow plants 38
Yorkshire fog 179
Yucca filamentosa 'Bright Edge'
  45
Yushania maculata 161

Z

zebra grass, see Miscanthus
  sinensis 'Zebrinus'

ACKNOWLEDGMENTS

The publisher would like to thank the following for their kind permission to reproduce their photographs:

(Key: a-above; b-below/bottom; c-centre; f-far; l-left; r-right; t-top)

6 Clive Nichols: Hampton Court 98 (bl). **Andrew Lawson:** Morville, Shropshire. Designer: Kathy Swift (br). **7 John Glover** (tl). **20 Dorling Kindersley:** RHS Wisley (crb). **22 The Garden Collection:** Derek Harris (l). **22-23 The Garden Collection:** Nicola Stocken Tomkins (c). **23 The Garden Collection:** Jonathan Buckley; Design: Anthony Goff (tr); Nicola Stocken Tomkins (br). **24 The Garden Collection:** Gary Rodgers; Designer: Christopher Bradley-Hole – Chelsea 2005 (tl); **Jonathan Buckley:** Design: Christopher Lloyd, Great Dixter (r). **Liz Eddison:** Designers: Beth Houlden and Louise Elliott – Chelsea 2005 (b). **25 The Garden Collection:** Nicola Stocken Tomkins (tl). **The Garden Collection:** Andrew Lawson; Sticky Wicket, Dorset (b). **28 Clive Nichols. 36 Harpur Garden Library:** Jerry Harpur / Bingerden, Netherlands (tr). **37 Clive Nichols:** Designer: Elisabeth Woodhouse (bl). **39 Clive Nichols:** Chiff Chaffs, Dorset (tl); Pettifers, Oxfordshire (tr). **Jo Whitworth:** Rob Whitworth, RHS Garden, Wisley (bl). **40 Jo Whitworth:** Sheila Chapman, Chelmsford, Essex. **51 FLPA:** Holt Studios International (fcrb). **Wildlife Matters:** (fcr). **58 Garden Picture Library. 80 Garden Picture Library:** John Glover. **100 Clive Nichols:** Beth Chatto Garden, Essex. **121 Andrew Lawson:** (tl). **Clive Nichols:** Little Coopers, Hampshire (tr). **122 Andrew Lawson:** (br). **123 John Glover:** (bl). **150 Clive Nichols:** Lady Farm, Somerset (b). **151 Clive Nichols:** The Old Vicarage, Norfolk. **156 Andrew Lawson. 157 John Glover:** Lois Brown, Pasadena, California (tr). **Clive Nichols:** Joe Swift (tl). **168 Dorling Kindersley:** Peter Anderson (tr). **Garden Picture Library:** Howard Rice (tr). **Photos Horticultural:** (tl). **176 FLPA:** Holt Studios International (bc). **Photos Horticultural:** (bl) (br). **177 Garden World Images:** (tc). **178 FLPA:** Holt Studios International (tr) (clb). **Photos Horticultural:** (cla). **179 Garden World Images:** (fcla). **Science Photo Library:** Pam Collins (bl); Maurice Nimmo (c). **186 Andrew Lawson. 194 Clive Nichols:** Lakemount, Cork, Eire. **195 Clive Nichols:** The Anchorage, Kent (b); White Windows, Hampshire (t). **210-211**

Dreamstime.com: Wiertn. **212 Andrew Lawson. 224 Marie O'Hara:** Planting Design: Paul Williams. **233 The Garden Collection:** Liz Eddison (tc). **250 Dorling Kindersley:** RHS Hampton Court Flower Show (tr). **255 Dorling Kindersley:** Alan Buckingham (c); RHS Hampton Court Flower Show (bc). **262 Dorling Kindersley:** RHS Chelsea Flower Show 2012 (tc). **263 Dorling Kindersley:** Peter Anderson (cb). **265 123RF.com:** xalanx (tr). **267 Dorling Kindersley:** Alan Buckingham (tl, br). **268 123RF.com:** Tatyana Aleksieva-Sabeva (bl). **271 Getty Images:** Stefano Stefani/Photodisc (tl). **272 Dorling Kindersley:** Alan Buckingham (bl). **285 Photos Horticultural. 302 Brian North:** (bl) (br). **314 Garden World Images:** (bl); Jacqui Dracup (bc); G. Kidd (br). **315 Garden World Images:** (tl); Jacqui Dracup (tc) (tr). **316 Garden World Images:** (bc). **Photos Horticultural:** (bl). **Premaphotos Wildlife:** Ken Preston-Mafham (br). **317 FLPA:** Holt Studios International (br). **Garden World Images:** Jacqui Dracup (bl) (bc). **318 FLPA:** Holt Studios International (b). **Garden World Images:** (cra) (crb); Jacqui Dracup (t); G. Kidd (c). **319 Garden World Images:** I. Anderson (tl); G. Kidd (tc) (bc) (bl) (cb) (clb). **322 FLPA:** Holt Studios International (ca). **Garden World Images:** (tc) (tr) (cra) (cb) (br); G. Kidd (bc) . **323 Dorling Kindersley:** Alan Buckingham (cla). **FLPA:** Holt Studios International (tl) (clb) (tc) (bc). **Garden World Images:** G. Kidd (ca) (bl); Jacqui Dracup (cb). **325 Photos Horticultural:** (tr).

All other images © Dorling Kindersley
For further information see: www.dkimages.com

Dorling Kindersley would like to thank: Andrew Halstead and Beatrice Henricot at RHS Wisley for their help with the Garden Doctor chapter, and the staff at the Royal Horticultural Society, especially Rae Spencer-Jones.

Design and editorial assistance: Zia Allaway, Murdo Culver, Elaine Hewson, Rachael Smith

Index: Vanessa Bird

Contributors Guy Barter, Alan Buckingham, Philip Clayton, Andi Clevely, Isabelle Van Groeningen, Lia Leendertz, Ian Spence, Alan Toogood, Daphne Vince Prue, Matthew Wilson

SECOND EDITION
Project Editor Amy Slack
Senior Designer Glenda Fisher
Senior Jacket Designer Nicola Powling
Jackets Co-ordinator Lucy Philpott
Production Editor David Almond
Production Controller Rebecca Parton
Managing Editors Ruth O'Rourke
Managing Art Editor Christine Keilty
Art Director Maxine Pedliham
Publishing Director Katie Cowan

DK DELHI
Managing Editor Soma B. Chowdhury
Senior Editor Janashree Singha
Assistant Editor Ankita Gupta
Senior DTP Designer Pushpak Tyagi
DTP Designers Anurag Trivedi, Satish Gaur
Pre-production Manager Sunil Sharma

This edition published in 2021
First published in Great Britain in 2008 by
Dorling Kindersley Limited
DK, One Embassy Gardens, 8 Viaduct Gardens, London, SW11 7BW

A CIP catalogue record for this book
is available from the British Library.
ISBN: 978-0-2414-5976-8

Printed and bound in China

For the curious

www.dk.com